A Workshop in Risk-Based Framing of Climate Impacts in the Northwest:

Implementing the National Climate Assessment Risk-Based Approach

Workshop Summary Authors
Meghan Dalton & Philip Mote
Oregon State University

Likelihood Synthesis Main Authors
Philip Mote & Peter Ruggiero
Oregon State University

Jeremy Littell, Dennis Lettenmaier, & Jan Newton
University of Washington

Sarah Shafer
U.S. Geological Survey

Jeffrey A. Hicke
University of Idaho

Workshop Facilitators
Kathie Dello, Josh Foster, Philip Mote, & John Stevenson
Oregon State University

March 2012

Table of Contents

A Workshop in Risk-Based Framing of Climate Impacts in the Northwest: Implementing the National Climate Assessment Risk-Based Approach

INTRODUCTION

The National Climate Assessment (NCA) is a report that summarizes and communicates the current understanding of climate change science and impacts in the United States. The Global Change Research Act of 1990, Section 106, requires a quadrennial national assessment on climate culminating in a NCA report that a) integrates, evaluates, and interprets the findings of the US Global Change Research Program and discusses the scientific uncertainties associated with such findings, b) analyzes the effects of global change on particular sectors, and c) analyzes current trends in global change, both human-induced and natural, and projects major trends for the subsequent 25 to 100 years (USGCRP).

The first NCA was released in 2000 and the second in 2009. To advance the NCA beyond a periodic report-writing activity, the 2013 report strives to build capacity for ongoing, comprehensive assessments by fostering partnerships with the public and private sectors, evaluating progress in adaptation and mitigation, identifying national indicators of change, providing web-based information to support decision making, and including new methods for documenting climate related risks and opportunities. To advance the conversation about climate impacts and adaptation research, the NCA has adopted a risk management framework for assessing climate impacts (USGCRP).

According to the Fourth Assessment Report of the Intergovernmental Panel on Climate Change (IPCC AR4), "responding to climate change involves an iterative risk management process that includes both adaptation and mitigation and takes into account climate change damages, co-benefits, sustainability, equity and attitudes to risk" (IPCC, 2007b). The NCA has adopted this framework to organize and prioritize key vulnerabilities according to the following criteria: risk (in terms of the likelihood and magnitude of consequences), timing, persistence, distributional aspects, potential for adaptation, and importance to those at stake.

The Northwest is primed to embrace this new perspective by building on over 15 years of high quality climate impacts and adaptation research in the region, and equally important, a sophisticated and well-educated community of manages and policymakers. Since the 2009 National Climate Assessment report, the Northwest has produced comprehensive state assessments[1] of the impacts of climate variability and change in Oregon and Washington, with much of the analysis and information extending to include Idaho as well. With such a wealth of work characterizing the current and future climate changes and

---

[1] Washington Climate Change Impacts Assessment produced by the Climate Impacts Group at the University of Washington in 2009; Oregon Climate Assessment Report produced by the Oregon Climate Change Research Institute at Oregon State University in 2010.

impacts in the region, the Northwest, led by the Climate Impacts Research Consortium (CIRC), is prepared to begin an assessment of climate risks region-wide.

To implement the NCA's risk-based framework, we[2] designed a workshop to bring together a breadth of experts from the region to engage in a discussion and assessment of climate risks moving beyond compiling a list of consequences to developing ways to rank the magnitude of consequences and prioritize risks and key vulnerabilities. The workshop builds on the climate risks identified in the Oregon Climate Change Adaptation Framework by surveying the participants and facilitating group discussions to compile consequences and begin development of a ranking system of these risks within the scope of the entire Northwest region. We describe the design of the workshop highlighting considerations of applying the risk-based framework to our region and lessons learned throughout the process. A summary of the outcomes of the workshop is presented for each climate risk considered.

WORKSHOP DESCRIPTION

The Northwest NCA Risk Framing Workshop was conceived as a way to bring together carefully selected experts from the region to engage in a discussion of climate impacts through a risk frame, separately assessing the likelihood and consequences. The objectives of this integrative workshop were three-fold: 1) discuss and rank the likelihood and consequences of climate risks to the northwest region, 2) provide opportunity to help inform the NW chapter of the NCA, and 3) build capacity for a long term, sustained regional assessment process. This workshop, held in Portland, Oregon, served as an initial step toward identifying key risks and vulnerabilities to highlight in the Northwest chapter.

We considered it important to include representatives from all sectors, affiliations, and states and communities within the region to ensure that the outcomes of this workshop are truly representative of the collective region. To that end, we asked participants a series of demographic questions. The workshop consisted of four main components: 1) introduction to risk-based framing of climate impacts; 2) a panel of experts presenting on the likelihood of eight climate risks; 3) an online, real-time survey collecting, from each participant, responses to questions about the consequences of those risks; and 4) breakout group discussions.

Demographics
Through a survey completed by 41 of the 48 workshop participants, we asked participants about their views on climate change, their affiliations, and expertise. All survey participants are climate conscious and savvy people with all participants indicating they were at least moderately concerned and moderately informed about climate change. About three-fourths were very concerned and about 80% were well informed about climate change.

---

[2] Meghan Dalton, Ginger Armbrust, Kathie Dello, Paul Fleming, Philip Mote, and TC Richmond

Most attendees were researchers, staff, or managers of their organizations. A few educators, directors, coordinators, advisors, and technical specialists were also in attendance. Slightly more than one-third of workshop attendees were from universities and slightly less than one-third from federal agencies. State agencies, local, county, and regional organizations, tribes, private companies, and an NGO were also represented at the workshop. For a list of organizations represented, see Appendix A. About one-third of attendees had been with their organization for over 10 years and two-thirds had been with their organization for 10 years or less. One-third had been with their organization for less than 5 years.

The main areas of academic training of the attendees were the biological, environmental and physical sciences. Several attendees were trained in engineering, social science, planning, economics, policy, and public health. The self-identified sectors of expertise represented include ecosystems, public health, environmental quality, climate and climate modeling, ecology, biology, hydrology and water resources, geology, infrastructure, tribal issues, coastal, agriculture, forestry, policy, and economics. Participants engage in activities such as restoration, spatial analysis, data management, impacts assessments, adaptation, climate policy, mitigation, conservation, communications, decision support, utility regulation and operation, computer modeling, reclamation studies, uncertainty and risk analysis, resource management, and land use planning. Representatives from Oregon, Washington, and Idaho were in attendance with geographic areas of expertise including both sides of the Cascade Range, Columbia, Snake, Yakima, and Deschutes River basins, Rocky Mountains, sagebrush steppe, coastal zones including the Puget Sound, and volcanic provinces.

Risk framework introduction[3]

To inform participants of the framework for this workshop, which is the NCA framework for incorporating iterative risk management in climate impact assessments, we invited two introductory speakers. One speaker was Meg Jones, of the Washington State Office of the Insurance Commissioner, who provided a look into the common use of a risk management framework within the insurance industry and compared and contrasted the application of risk management to climate impacts. Dr. Gary Yohe, Vice-Chair of the NCA Development and Advisory Committee, had developed the NCA guidance for incorporating the risk-based framework; Dr. Yohe presented (remotely via speakerphone) the NCA guidance on risk framework and how to handle uncertainty.

A working definition of risk is the product of likelihood and consequence. Likelihood depends on the variability, forcing, and sensitivity of the climate and the consequence can be assessed through a variety of metrics from physical impacts to vulnerability. The importance of risk framing is outlined in the Summary for Policymakers of the IPCC Fourth Assessment Report synthesis (IPCC, 2007b) as an iterative process including both adaptation and mitigation while taking into account not only damages, but co-benefits of climate change in addition to sustainability, equity and attitudes to risk. Subsequently, this framework has been adopted by all four National Research Council Panel Reports of

---

[3] A summary of Gary Yohe's presentation

America's Climate Choices (NRC, 2010a; NRC, 2010b; NRC, 2011), the New York City Panel on Climate Change (NPCC, 2010), the draft Adaptation Plan for the United States, and the Department of Defense.

The NCA strives to sustain an "ongoing analysis of scientific understanding of the climate change impacts, risk, and vulnerability that is relevant to a wide range of decisions and policies" (NCA Strategic Plan). The NCA Regional Strategy encourages technical input teams to use an integrated, risk-based approach to the extent possible to characterize key vulnerabilities. Criteria for judging key vulnerabilities, according to the IPCC fourth assessment report, include: magnitude, timing, persistence/reversibility, potential for adaptation, distributional aspects, likelihood, and importance to relevant actors.

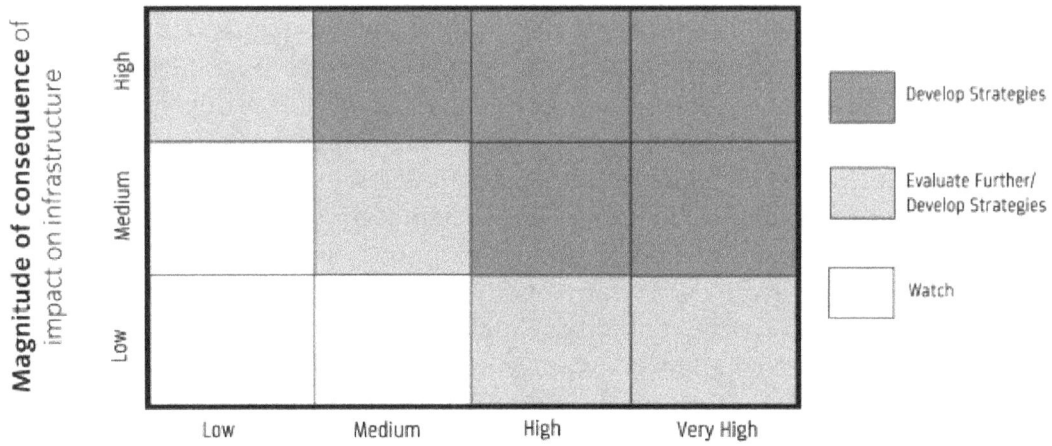

Figure 1 Risk matrix used and developed by the New York Panel on Climate Change, and adopted by the National Climate Assessment.

The risk matrix can be used as a tool to organize thoughts around key vulnerabilities and risks (NPCC, 2010). The estimation of risks can be based on quantitative estimates or qualitative representations of likelihood and consequences. According to Dr. Yohe's presentation, these risk matrices "are most effective when various stakeholders defend their portraits over time before one another". For this purpose, we gathered representative stakeholders from the region to engage in group discussions.

Dr. Yohe described some guidance for assessing and calibrating consequences and likelihood, found in his draft NCA guidance document *Adopting a Risk-Based Approach Informed by "Key Vulnerabilities" in the NCA Process*. For consequences, define a metric (e.g. economic, human health) and describe the rationale for the approach selected, address the sensitivity of consequences to the influence of multiple stresses, move beyond physical impacts to consider vulnerability as a function of exposure, sensitivity, and adaptive capacity. When assessing likelihood, describe the process and source from which the likelihood judgment emerged and use the language in the NCA's uncertainty guidance. An underlying, yet important caveat is that a traceable account of sources of estimations of likelihood or consequence, whether qualitative or quantitative, must be provided and as complete as possible.

Likelihood can be compared across risks if the scales are derived from the same or comparable climate models and drivers. According to Dr. Yohe's presentation, "consequence scales are often expressed in different metrics, so stakeholder input is required to develop comparable judgments about the magnitude of the consequences", in other words, publically defending judgments. With qualitative or quantitative rankings of likelihood and consequences of several climate risks, ensuring a complete traceable account of information and judgment, risks can be prioritized based on where they fall on the risk matrix. The NCA will focus not only on high likelihood – high consequence risks, but also low likelihood risks that could carry high consequence. For the likelihood assessment portion of ranking climate risks, we chose to enlist a panel of experts to formulate an initial determination of the likelihood of each climate risk.

Likelihood panel

The purpose of the short panel presentations on likelihood was to introduce each risk to the workshop participants in terms of drivers, geographic area affected, time frame, and most importantly, the likelihood and/or confidence of occurrence. Because the assessment of likelihood or probability involves careful expert judgment and evaluation of recent literature, we selected experts in the field to prepare a brief synthesis for each risk. We asked of each person that, prior to the workshop, they determine the best estimate and range (5% and 95% confidence bounds) for the likelihood of this impact occurring mid-21st century relative to the late 20th century, or for whatever time scales there exists literature). In addition, we asked that they prepare a written synthesis with citations articulating their reasoning and evidence for the likelihood determined. These syntheses are included in the likelihood subsection of each risk summary below.

The panel of likelihood experts then presented short summary of each of the climate risks to rank during the remainder of the workshop. The identified risks were drawn from the Oregon Climate Change Adaptation Framework (2010) and Washington Climate Change Impacts Assessment (2009). The significant aspects of each risk were presented, including: drivers, geographic area affected, time frame, and in particular, the likelihood of occurrence. Presenters provided the preliminary likelihood ranking for risk prioritization. After each five-minute presentation, workshop participants were then asked to answer survey questions regarding the consequences and magnitude of those consequences for the risk just presented.

Survey

Given the likelihood of the risk just presented, workshop participants were asked to complete a survey in which they were asked to list all the consequences of this risk in their area of expertise, and answer a series of rating questions regarding the magnitude of consequences according to a set of pre-determined metrics. The questions were exactly the same for each risk. See Appendix B for the survey questions used. The goal of the survey was to get people thinking individually about consequences before engaging in group discussion, provide useful data for analysis and synthesis after the workshop, allow everyone to provide input on each climate risk, and provide preliminary data for later use within the afternoon breakout sessions.

The survey rating questions provided a preliminary set of metrics to assess the magnitude of consequences. The survey asked: "To the best of your knowledge, how would you rate the consequences of this risk in the following criteria": Human Health and Welfare (in terms of human population affected, human mortality, & health quality), Natural Environment (in terms of biodiversity and geographic area affected), Built Environment (in terms of public services and infrastructure), Economy (in terms of cost). The rating criteria and scales under Human Health and Welfare and Economy were borrowed and adapted from the City of Atlanta's climate change vulnerability and risk assessment (Morsch and Saterson, 2010). The biodiversity rating criterion was borrowed from NPCC (2010), geographic area affected criterion was adapted from Snover et al. (2007), and the public services and infrastructure criterion was adapted from NPCC (2010).

The answer format was multiple-choice, with one choice per question allowed, and included space to provide rationale for the answer selected. The response options for each question were descriptive and included degrees of positive and negative options. For example, the response options for the rating of consequences in terms of human mortality were: Decrease in Mortality, Sign Uncertain, Increase in Mortality, Don't Know, and N/A. By including Don't Know and NA response options, we provided a way for survey participants to actively opt out of the question.

An initial draft of survey questions was circulated among the workshop facilitators for comment and revision. We would recommend to other groups considering administering a similar survey to involve a social scientist and/or communications expert in the development of the survey. While many positive comments were received on the survey technique as a way to capture opinions from local experts, there was some concern about the appropriate knowledge and time for providing answers to some of the rating questions. Some questions, particularly the question involving economic cost, were said to be simply unanswerable at this time or that answers depended on various factors.

We chose to administer the survey through SurveyMonkey, a provider of web-based survey tools. While we considered using a hand-held audience response system, the SurveyMonkey software allowed an opportunity for participants to provide open-ended comments and the workshop facilitators to quickly compile and download survey reports on a subset of survey questions for use during the workshop. Quantitative and some qualitative survey data were compiled during a break and distributed for use during afternoon breakout sessions. See Appendix D for an example of the survey reports handed out to the breakout groups. The idea was that expert judgment within the groups would be practiced to refine and/or override preliminary rankings, organize consequences, and produce coherent rationale for the relative ranking of the likelihood of occurrence and magnitude of consequences of climate risks.

From the panel presentations, the workshop participants heard an introduction to each climate change risk and, where possible, a preliminary judgment of the likelihood. The surveys provided an opportunity for workshop participants to provide their input on all risks. The goal of the breakout sessions was to develop a preliminary ranking of the magnitude of consequences and along with rationale. The final plenary discussion was designed to engage the entire group in a discussion of all the risks together on a common risk matrix.

There were two breakout group sessions during the afternoon. Each breakout session consisted of four smaller groups, one for each risk. Each person selected a total of two risk groups, one during each breakout session. We did not assign people to groups, but let everyone choose which breakout group to participate in. This resulted in an uneven distribution among all the groups, but every group had at least four people. Each group was led by a facilitator whose role was to ensure the documentation of the process, group results, and other useful notes, and to keep the group on track with the charges. The charges to the breakout groups were:

1. Identify and describe populations and areas vulnerable to your group's climate risk. Why and how are they vulnerable?
2. What are specific risks to these vulnerable populations and/or areas? List, describe and categorize the consequences conveying why the risk(s) are a source of societal concern.
3. Develop a ranking scheme and rationale for these consequences that is appropriate to your group's risk(s), but can also be comparable with other risks. Use or refine the provided ranking criteria from the surveys, or develop a new one. Qualitatively rate the magnitude/severity of the consequences (i.e. modest to severe; low to high). Provide written rationale for rating.
4. Discuss sensitivity of consequences to the magnitude of climate change and how the magnitude of consequences might change over time. What factors may influence the sensitivity of consequences?
5. Discuss the ranking of the likelihood of occurrence of your group's risk(s). Do you agree with the preliminary ranking? How would you modify it? Provide written rationale.

After each breakout session, the groups reported to the plenary group and had time for questions, comments, and discussion. We did not have time for a final plenary discussion, but the discussions during the breakout group reports brought up several challenges and points of consideration of the risk-based framework. While much of the breakout group discussions were around the first two charges, some groups moved into the third charge ranking consequences. Along with the likelihood presentations and survey results, the breakout group discussions are summarized in the following section, which contains a summary of each of the climate risks considered during this workshop.

RISK SUMMARIES

The climate risks considered during this workshop were borrowed and adapted from the risks identified in the Oregon Climate Change Adaptation Framework (2010), with consideration also of the impacts discussed in the Washington Climate Change Impacts Assessment (2009). The eight risks are: 1) extreme heat events, 2) changes in hydrology and water resources, 3) wildfires, 4) ocean temperature and chemistry changes, 5) sea level rise and coastal hazards, 6) shifts in distributions of plants and animals, 7) increase in invasive species, pests and diseases, and 8) extreme precipitation and flooding. The risk summaries were developed from three workshop materials including the likelihood syntheses and presentations prepared prior to the workshop, results from a written survey conducted during the workshop, and notes from each of the breakout groups. Each risk summary contains the following subsections: likelihood, consequences, and summary of survey rating questions. The following subsections are also included, if input was provided: case study examples, identified gaps in knowledge, special considerations, and references.

The likelihood section consists of the synthesis or outline written by those who gave (or contributed to) respective presentations at the workshop. Some syntheses were modified to extend geographic coverage or provide more background. Most include a summary of the climate drivers, description of the manifestation of the risk in the Northwest, and discussion of the likelihood and confidence of projected future changes where supportive literature exists. In the consequences section, consequences are summarized in a table organized by five categories: Natural Systems, Managed Systems, Built Environment, Economy, and Human Society. The information in the table is drawn from the rating question comments and the open-ended question asking participants to list all consequences in their respective area of expertise, and from the breakout group notes. Positive consequences and opportunities are highlighted in bold within the table. Citations are included wherever they were provided. Some breakout groups agreed on a qualitative initial rating of certain consequences. Those ratings are reflected in italicized parentheses.

The section containing a summary of the survey rating questions is intended to help inform and provide rationale for pertinent metrics of use to rate the magnitude of consequences. The quantitative results are presented along with a summary highlighting many of the rating question comments. The section of case study examples contains an annotated list of pertinent case studies mentioned or provided by workshop participants either through the survey or during breakout discussions. Likewise, the gaps in knowledge and special considerations sections contain information brought forth through workshop activities. References from each section are collected at the end of each risk summary.

While the definition of mean (or average) values is straightforward, approaches to defining extremes vary considerably depending in part on application. For example, high temperature extremes could be defined by the warmest day of the year, or by a quantity that may have more relevance to impacts on human health: average minimum temperature over three consecutive days. Computing trends or long-term changes in extremes involves a tradeoff between obtaining enough events for robust statistics, and having the events be extreme enough to be consequential. It is common to achieve robust statistics in part by aggregating results over a wide area, for example the Northwest. For future changes, we primarily use data from 9 simulations in the North American Regional Climate Change Assessment Project (NARCCAP, Mearns et al. 2007), at roughly 50 km resolution, for 2041-2070 relative to 1971-2000.

Observed changes in heat waves in Washington and Oregon: Bumbaco et al (unpublished manuscript) examined heat waves using a definition of 3 consecutive daytime (or nighttime) temperatures above the 99th percentile for May-September, after aggregating over sub-state spatial domains. They found no significant change in heat waves expressed as daytime maximum temperatures, but a large increase since 1980 of heat waves expressed as excessively high nighttime minimum temperatures. The data set had been adjusted for instrumental changes, station moves, and urban influence.

Projected changes: models are unanimous that measures of heat extremes will increase and measures of cold extremes will decrease (Table 1).

Table 1. The mean and standard deviation of changes in selected temperature variables for the NARCCAP simulations. Mean changes from the CMIP3 statistically downscaled analyses are also shown for comparison. (Kunkel et al, 2012).

Variable Name	NARCCAP Mean Change	NARCCAP St. Dev. of Change	Statistically-Downscaled Mean
Freeze-free period	+35 days	6 days	+30 days
#days Tmax > 90°F	+8 days	7 days	+19 days
#days Tmax > 95°F	+5 days	7 days	+11 days
#days Tmax > 100°F	+3 days	6 days	+4 days
#days Tmin < 32°F	-35 days	6 days	-25 days
#days Tmin < 10°F	-15 days	7 days	-9 days
#days Tmin < 0°F	-8 days	5 days	-4 days
Max run days > 95°F	+134%	206%	+251%
Max run days > 100°F	+163%	307%	+501%
Heating degree days	-15%	2%	-16%
Cooling degree days	+105%	98%	+157%

---

[4] Authors: Philip Mote, Kathie Dello, Kenneth E. Kunkel, & Laura E. Stevens

Growing degree days (base 50°F)	+51%	14%	+51%

Consequences:

Category	Consequences (**bold** denotes a positive consequence)
Natural Systems *(medium high)*	• Decreased snowpack and summer water supply and changes in runoff timing • Cold water fish kills, fisheries extinction • Salmon migration timing altered • Local extirpation of tree species • Loss of soil infiltration capacity with prolonged drought/heat wave • Increase carbon loss to atmosphere from fire and taiga • Increase stressors on species & habitats • Species range shifts, community fragmentation, loss • Changes in water flow to estuaries (e.g. nutrient delivery, stratification) • Decline in water quality of steams & estuaries • **Some trees do well with high vapor pressure deficit (are relatively insensitive to high temperatures)**

Managed Systems (e.g. Agriculture, Forestry) *(medium low)*	• Agriculture (maximum temperature more important than number or duration of events):
	  o Crops fail to harden over winter
	  o Reduced yields when plants/animals suffer heat stress
	    ■ E.g. Wheat/maize – temps > 30C; Milk production decrease w/ each degree above ideal; Corn – threshold of 28C
	  o Increased water needs, decreased water availability
	• Forests:
	  o Tree mortality (already occurring)
	  o Increased disturbances (including risk of extreme drought)

Built Environment	• Material stress for roads, bridges, dams from extreme heat (especially concrete) • Could shorten lifetime of infrastructure (e.g. dams)
Economy	• Loss in agricultural revenues (especially fruit & wine grape growers) depends on water supply; Yakima valley is especially vulnerable. • Food shortage leading to rising prices (e.g. wheat/grain) • Sharp increase in demand on water supplies during heat events (e.g. agricultural and municipal) • Energy grid failures (i.e. outages) during peak summer electricity demands (Hamlet et al. 2010; NWPCC, 2010). • Altered economic expectation in forest management

	• Losses to recreational ski areas (winter warm spell can kill a ski season – e.g. winter 2011) • **Increased wind energy production during inland heat waves**

Human Society
(low)
	• Increase in heat-related acute & long-term health impacts & mortality (e.g. heat rash, heat syncope, heat exhaustion, heat stroke)
	   ○ 33 hospitalizations per year related to heat in Oregon (Oregon Health Authority)
	   ○ 1.8 heat related deaths per year in Oregon (Oregon Health Authority)
	• Effect on respiratory health from poor air quality from air stagnation and forest fires during heat events
	• Increase vector-borne disease
	• Impacts to culturally important species
	• Shifts in growing season & access to species for subsistence & economic needs
	• **Some positive interactions with respiratory ailments**
	• **Potential for fewer cold season deaths**

Summary of Survey Rating Questions

Human Population Affected: In terms of the portion of human population affected by increasing extreme heat events, the number of responses was about the same for each response, but with a slightly higher response incidence of 'Most'. Humans may be indirectly affected by water supply, air quality issues, food and agriculture, and associated fires. Direct consequences of increasing extreme heat events will likely vary depending on vulnerability with the elderly, living in low socioeconomic systems as the most vulnerable population. The magnitude of consequences is also moderated by adaptive capacity and depends on relative affluence and access to AC services. The regions with most exposure to increased extreme heat events will likely be those that are already hot and adapted to heat (e.g. interior Columbia Basin).

Human Mortality: Fifty-five percent of respondents indicated an increase in human mortality from increasing extreme heat events. The remaining 45% were uncertain of the sign or didn't know. Increasing extreme heat events will impact people in urban areas and especially the elderly, children, sick, and poor populations. Evidence of increased mortality from increasing extreme heat events is already observed (Jackson et al. 2010), and extreme overnight lows have shown increasing mortality in places not acclimated to such extremes. On the other hand, there could be a decrease in deaths cause by cold temperatures and freezing.

Human Health Quality: Fifty-five percent of respondents indicated reduced general health quality while the remaining 45% were uncertain of the sign or didn't know. The reduction of health quality, including exacerbated asthma, strokes, discomfort, and heat stress, may depend on mitigation strategies. Some indirect effects relating to human health quality may be increased atmospheric particulates from increased fire or increased disease from contaminated shellfish.

Biodiversity: About 60% indicated either unfavorable or severe reduction in biodiversity. Over a quarter indicated they didn't know. The risk of increasing extreme heat events is generally unfavorable for some species, but favorable for other species. A shift in distributions of species is expected with some species displacing others. Niche species are especially vulnerable, for example, native cold water fish. The degree of consequences will depend on the individual species' capacity to adapt.

Geographic area affected: Ninety-five percent of respondents indicated that either a moderate portion or most of the geographic region would be affected by increases in extreme heat events. Warming will impact all areas of the northwest region, but more so in the eastern side of Oregon and Washington and especially southern Idaho (Kunkel et al, in prep).

Built Environment: Approximately two-thirds of respondents indicated a moderate disruption in public services and infrastructure from increasing extreme heat events. Consequences to the built environment depend on timing and adaptive capacity. Increasing extreme heat events could change material performance and durability in high temperature regimes and could create higher energy and air conditioning demands. However, there may be power shortages, especially during summer, when there will likely be lower stream flow and hydropower generation.

Economy: Economic industries potentially affected by increasing extreme heat events include salmon fisheries, agriculture, hydropower, and forest systems. The extent of economic impact depends on adaptive management strategies, time scale, and sign of impacts on natural resources. For example, forest productivity may increase from longer growing periods.

Case Study Examples:
Partially as a result of increasing temperatures and extreme heat events, the peak in electricity load is moving from winter heating to summer air conditioning. A few locations in the Northwest already have summer peaks in electricity load. (Hamlet et al. 2010; NWPCC, 2010).

Identified Gaps in Knowledge:
Important impacts pertaining to changes in agriculture and forest productivity due to extreme heat events are not explicitly represented.

Special Considerations:
The magnitude of the consequences of extreme heat events is strongly affected by adaptive capacity, like the availability and use of air conditioning during heat wave events. The risk of increases in extreme heat events is closely related to several other risks including increased chance of wildfire, increased risk of pathogens, diseases, and cold-tolerant species, and increased risk of drought.

The aggregate position on the risk matrix that this breakout group came up with is a tear drop shape with the bulbous part in the low to medium consequences and the narrow source of the tear would be in the high consequence box, all with very high likelihood. This reflects the thought that most of the consequences would be low to medium with a few high consequences.

References

Bumbaco, K., K. Dello, and N. Bond. Historical Analysis of Pacific Northwest Heat Waves. In preparation.

Hamlet, A.F., S.Y. Lee, K.E.B. Mickelson, and M.M. Elsner. 2010. Effects of projected climate change on energy supply and demand in the Pacific Northwest and Washington State. Climatic Change 102(1-2): 103-128, doi: 10.1007/s10584-010-9857-y.

Jackson, J. Elizabeth, M. G. Yost, C. Karr, C. Fitzpatrick, B. K. Lamb, S. H. Chung, J. Chen, J. Avise, R. A. Rosenblatt, R. A. Fenske, 2010. Chapter 10: Public health impacts of climatic change in Washington State: Projected mortality risks due to heat events and air pollution. Washington State Climate Change Impacts Assessment: evaluating Washington's future in a changing climate.

Kunkel, K. E., L. E. Stevens, S. E. Stevens, E. Janssen, and K. T. Redmond. Climate of the Northwest U.S. In preparation.

Mearns, L.O., et al., 2007, updated 2011. The North American Regional Climate Change Assessment Program dataset, National Center for Atmospheric Research Earth System Grid data portal, Boulder, CO. Data downloaded 2012-02-24. [http://www.earthsystemgrid.org/project/NARCCAP.html].

Northwest Power and Conservation Council (NWPCC), 2010. The sixth Northwest conservation and electric power plan, appendix L, Climate Change and Power Planning. [http://www.nwcouncil.org/energy/powerplan/6/default.htm].

Likelihood:[5]

Projected future changes in hydrology and stream flow are sensitive to projected future changes in temperature and precipitation across the region. Some of the hydrological issues of climate change facing the Northwest include: changes in seasonal cycle of runoff, especially in transient and snow-dominated basins; changes in annual runoff; changes in precipitation extremes, especially as it affects urban runoff; and changes in extreme floods.

Projected changes in seasonal cycle streamflow are different for the types of watersheds in the Northwest, which are rain-dominated, snowmelt-dominated, and transient (mixed rain and snowmelt). A typical hydrograph of a rain-dominated watershed shows a peak in streamflow during the cold season, when most of the precipitation falls. Streamflow in snowmelt-dominated watersheds typically peak in late spring or early summer as the snowpack melts. Transient watersheds, influenced by both rain and snowmelt, typically experience a winter streamflow peak, and a late spring, early summer peak. (Elsner et al. 2010). With projections of warming temperatures and increasing cool season precipitation, snowpack is expected to decline as more cool season precipitation falls as rain rather than snow. An implication for snowmelt-dominated watersheds is that the timing of streamflow may shift to resemble transient watersheds. Transient watersheds are especially sensitive to changes in temperature and precipitation, and the timing of streamflow may shift to resemble that of rain-dominated watersheds. The timing of streamflow in rain-dominated watersheds isn't expected to change, but the cool season peak may increase. Changing timing of streamflow has implications for water management in the region. (Elsner et al. 2010; Chang and Jones, 2010).

Elevation is an important factor when determining the change in basin runoff because it affects how much precipitation falls as snow in the winter and the rate snowpack melts in the spring. At higher elevations, changes in winter temperature are more important than winter precipitation for changes in winter runoff; changes in winter precipitation become more important at lower elevations (Chang and Jones, 2010). In the state of Washington, composite changes in annual and cool season runoff are projected to increase while warm season runoff is projected to decrease (Elsner et al. 2010). In the Willamette River basin in Oregon, projections of ensemble mean changes in runoff are shown to decrease in summer and increase in winter. There is high uncertainty in projected future runoff, particularly in places where groundwater is a big factor in the seasonal water cycle, such as the High Cascades (Change and Jung, 2010).

An increase in precipitation extremes is projected for future climate change. In a case study of Portland, Oregon, Chang et al. (2010) present evidence for increase in frequency of storm events likely leading to more common flooding of road cross-sections that already experience such flooding. Mirsha and Lettenmaier (2011) analyzed trends in annual precipitation maxima in large U.S. urban areas over 1950-2009, and found significant

---

[5] A summary of Dennis Lettenmaier's presentation with additional references

increasing trends in a majority of urban areas, but because of the topographical complexity, trends in urban areas in the northwest were more ambiguous.

Many extreme rainfall events and resulting flooding in the Northwest are associated with atmospheric river events. Ralph et al. (2011) present some examples of recent atmospheric rivers that produced extreme rainfall and flooding. It is possible that atmospheric river events may increase in the future brining more extreme floods. Projections of 21st century climate suggest more winters with many atmospheric rivers landing on the Central California coast and fewer winters with few land falling atmospheric rivers (Dettinger et al. 2009).

Consequences:

Category	Consequences (**bold** denotes a positive consequence)
Natural Systems	• Impacts on salmon & fisheries from altered timing (e.g. loss of juveniles from winter scouring floods, summer droughts affecting migrating/spawning adults) – *(high)* • Impacts to other at-risk/ESA species: salmon, trout, bulltrout, snails *(high)* • Water quality implications (streams & estuaries) • Increased stream temperature (Isaak et al. 2010) • Increased stream channelization *(low medium)* • Increased TDML to Puget Sound (nutrients, DOC, contaminants) • Habitat/refugia modification (e.g. riparian wetland) *(medium high)* • Mortality of forests from drought stress & low soil moisture *(medium)* • Changes in plant species composition (e.g. affect pollination ecology of invertebrates & wildlife in high elevation forests & meadows) • More frequent toxic red tide due to low stream flow (e.g. Puget Sound/Skagit river study, Moore et al. 2011) • Source mentioned: Crozier, 2011; Tillmann and Siemann, 2011a;
Managed Systems (e.g. Agriculture, Forestry)	• Yield reduction • Shift in agricultural crop types & methods • Impacts from droughts • Conflicts regarding water availability & allocation issues in summer (e.g. reduced water supply for irrigated agriculture *(high)*, urban *(medium)*, hydropower, forest ecosystems)
Built Environment	• Water supply infrastructure & operation/management (including municipal & storm water *(urban - high)*, Columbia R federal dam system, hydropower, reservoir capacity *(medium high)*) (see U.S. Bureau of Reclamation report)

	• Increased flood risk in some zones => structure damage
Economy	• Increased cost of electricity in summer/fall from lower electricity/hydropower production; increased demand (especially for agricultural growers) – *(medium high)* • Economic impacts from impacts on salmon & fisheries • Changes to current operation, infrastructure, & policies related to in-stream biology (e.g. Columbia River Treaty – *(high)*)
Human Society	• Impacts to tribal culture & subsistence from impacts to salmon & fisheries • Winter recreation impacted *(low)* • Decreased water quality (e.g. concentrated pollutants, increase bacteria, water borne disease *(low medium)*)

Summary of Survey Rating Questions:

Human Population Affected: About three-fourths of respondents indicated most or about half of the region's human population would be affected by changing hydrology and water resources. Many people will be affected to some extent in the Columbia Basin through impacts to water supply, hydropower, agriculture, flooding, recreation, and fisheries for example. "Tribal communities are particularly susceptible because of the cultural and economic importance that they place on salmon and other cold-water fish populations, which may be diminished or extirpated due to changes to hydrology in the Columbia Basin."

Human Mortality: Most (86.8%) of the respondents selected an answer that indicated no sign of change for human mortality (Don't Know, NA, or Sign Uncertain). The magnitude of the impact of hydrological change may depend on the degree of water scarcity and community adaptation strategies. Only five respondents (13.2%) indicated an increase in human mortality due to changing hydrology and water resources.

Human Health Quality: About one-third of respondents indicated reduced general health quality while roughly 60% chose an answer with no indication of the sign of change in human health quality (Don't Know, NA, or Sign Uncertain). Of those who indicated reduced general health quality, the reasons given include health effects of water shortage for drinking and agriculture and the effect of the availability of salmon as an important nutritional protein source for tribal communities. The risk of changing hydrology and water resources is an example of how the magnitude of consequences depends on community mitigation and adaptation strategies.

Biodiversity: Over half (57.9%) of respondents indicated that changing hydrology and water resources would be unfavorable for biodiversity, and some others (13.2%) indicated a severe reduction in biodiversity. Fresh water impacts, including reduction and timing of stream flow and temperature changes, on aquatic species and habitats will likely be unfavorable for biodiversity. Native species may be stressed and possibly extirpated and replaced by invasive species. Flooding and increased erosion may affect spawning and rearing habitats. Terrestrial systems may see an increase in water stress. About one-quarter of respondents indicated they didn't know.

Geographic area affected: About 95% of respondents indicated that either a moderate portion or most of the region would be affected by changing hydrology and water supply. Twice as many indicated that most of the region would be affected versus a moderate portion affected. A few comments indicated that all regions are impacted by changes in the hydrological cycle. Some respondents felt that aquatic ecosystems would be sensitive to changes in seasonal flow patterns, and terrestrial ecosystems would be sensitive to loss of snowpack and soil moisture. The east side of the region, that is, east of the Cascades Mountains, may be most affected.

Built Environment: Just less than one half of respondents indicated moderate disruptions, and one-third indicated major disruption of public services and infrastructure from changing hydrology and water resources. The built environment including hydropower, dam operation, agriculture, and municipal water supply will likely be affected. Adaptation strategies involving reorganizing dam operations and water efficiency may buffer the affect on water supply infrastructure.

Economy: Almost two-thirds of respondents indicated medium or high economic loss due to changing hydrology and water supply. About one-quarter of respondents indicated they didn't know. The major economic activities potentially affected are agriculture, forestry, fisheries (e.g. mitigation costs for salmon), recreation, and hydropower (e.g. reallocation of water, increase cost of energy, loss revenue). Past events that are similar to those expected in a warming world need to be examined in terms of their economic impact.

Case Study Examples:
More frequent toxic red tide in the Puget Sound partially due to low stream flow from the Skagit River (Moore et al. 2011).

Special Considerations:
The consequences of changing hydrology and stream flow depend on the watershed and stream reach according to work done by Mantua et al. (2009) and Hamlet et al. Related risks include increased wildfire risk, increased diseases, and increased flooding in some zones.

References:
Chang, H., and J. Jones, 2010. Chapter 3: Climate Change and Freshwater Resources in Oregon. Oregon Climate Assessment Report, K.D. Dello and P.W. Mote (eds). College of Oceanic and Atmospheric Sciences, Oregon State University, Corvallis, OR.
Chang, H., and I.-W. Jung, 2010. Spatial and temporal changes in runoff caused by climate change in a complex large river basin in Oregon. Journal of Hydrology, 388, pp. 186-207.
Chang, H. Lafrenz, M. Jung, I-W., Figliozzi, M., Platman, D. and Pederson, C. (2010a) Potential impacts of climate change on flood-induced travel disruption: A case study of Portland in Oregon, USA. Annals of the Association of American Geographers

100(4): 938-952

Crozier, L., 2011. Literature review for 2010 citation for BIOP: Biological effects of climate change. Northwest Fisheries Science Center, NOAA-Fisheries, August 2011.

Dettinger, M.D., Hidalgo, H., Das, T., Cayan, D., and Knowles, N., 2009. Projections of potential flood regime changes in California: California Energy Commission Report CEC-500-2009-050-D, 68 p.

Elsner, MM, L Cuo, N Voisin, JS Deems, AF Hamlet, JA Vano, KEB Mickelson, SY Lee, and DP Lettenmaier, 2010. Implications of 21st century climate change for the hydrology of Washington State. Climatic Change, 102(1-2): 225-260, doi: 10.1007/s10584-010-9855-0.

Isaak, Daniel J., Charles H. Luce, Bruce E. Rieman, David E. Nagel, Erin E. Peterson, Dona L. Horan, Sharon Parkes, and Gwynne L. Chandler. 2010. Effects of climate change and wildfire on stream temperatures and salmonid thermal habitat in a mountain river network. Ecological Applications 20:1350–1371. doi: http://dx.doi.org/10.1890/09-0822.1

Mantua, N., I. Tohver, and A. Hamlet, 2009. Chapter 6: Impacts of Climate Change on Key Aspects of Freshwater Salmon Habitat in Washington State. The Washington Climate Change Impacts Assessment, M. McGuire Elsner, J. Littell, and L Whitely Binder (eds). Center for Science in the Earth System, Joint Institute for the Study of the Atmosphere and Oceans, University of Washington, Seattle, Washington. Available at: http://www.cses.washington.edu/db/pdf/wacciareport681.pdf

Moore, Stephanie K., Nathan J Mantua, Eric P. Salathe Jr. 2011. Past trends and future scenarios for environmental conditions favoring the accumulation of paralytic shellfish toxins in Puget Sound shellfish. Harmful Algae, Vol. 10, Issue 5, 521-529. Doi: http://dx.doi.org/10.1016/j.hal.2011.04.004.

Ralph, F.M., P.J. Neiman, G.N. Kiladis, K. Weickmann, and D.W. Reynolds, 2011. A Multiscale Observational Case Study of a Pacific Atmospheric River Exhibiting Tropical–Extratropical Connections and a Mesoscale Frontal Wave. *Mon. Wea. Rev.*, **139**, 1169–1189.doi: http://dx.doi.org/10.1175/2010MWR3596.1

Tillmann, P. and D. Siemann, 2011a. Climate Change Effects and Adaptation Approaches in Freshwater Aquatic and Riparian Ecosystems of the North Pacific Landscape Conservation Cooperative Region: A Compilation of Scientific Literature. Phase 1 Draft Final Report. National Wildlife Federation – Pacific Region, Seattle, WA. August 2011.

U.S. Bureau of Reclamation, Climate and Hydrology Datasets for Use in the River Management Joint Operating Committee (RMJOC) Agencies' Longer-Term Planning Studies, Part IV–Summary Report, September 2011. http://www.usbr.gov/pn/programs/climatechange/reports/finalpartIV-0916.pdf.

Climate change impacts on fire risk in the Pacific Northwest

For consistency with the NCA Forests Technical Input chapter (Peterson and Vose, in prep), fire risk is defined separately from fire hazard, which is often confused with risk. Peterson and Littell use Hardy's definition therein: "Fire hazard is the structure, condition, and arrangement of a fuelbed as they affect its potential for flammability and energy release. Fire risk is the probability that a fire will ignite, spread, and potentially affect one or more resources valued by people (Hardy, 2005)." The "best estimate" of risk likelihood for changes in fire regime components (fire frequency, area burned, severity, and intensity) in the Pacific Northwest varies (1) with the fire regime component in question and (2) within the region according to how fire risk and fire hazard vary with fuels in different vegetation types that respond to climate in different ways. For the region as a whole, much more information on the effects of climate change on the area of fire is available than for changes in frequency, severity, or intensity.

Climate-driven increases in regional area burned by fire are very likely by mid-century (2040s relative to late 20th century), but more likely in forested ecosystems than in non-forested systems. Region wide, the probability of exceeding the 95% quantile area burned for the period 1916–2006 increases from 0.05 to 0.48 by the 2080s (Littell et al. 2010). The probability of exceeding the late 20th century historical (1980-2006) 95%ile under expected future hydroclimate (Elsner et al. 2010; Mote and Salathé, 2010, for A1B and B1 ensemble averages of 20 and 19 GCM realizations for the 2020s, 2040s, and 2080s) is a range of 0 to 30% for non-forested systems, 1 to 19% for the western Cascades, and 0 to 76% for the eastern Cascades, Blue Mountains, and Okanogan Highlands (after Littell et al. 2010). No statistical relationships could be constructed for the Oregon Coast Range and Olympic Mountains, though climate effects on fire in those ecosystems are evident in the more recent paleoecological record (Henderson et al. 1989) and the consequences of rare events are extreme (>0.5 million ha burned in single events).

The "natural" relationships between climate and fire are dependent on existing forest fuels, ignitions, and the climatic characteristics that facilitate or limit fire. On longer time scales, climate also affects fuels through forest biogeography and vegetation productivity. These relationships are evident from lake settlement charcoal and pollen studies from prior millennia (e.g. Power et al. 2008, Marlon et al. 2008). In the last century, ignitions and modified vegetation are virtually ubiquitous with human land use, so how climate affects the components of fire regimes (fire area, fire size, severity, frequency, intensity) in forests depends on the amount, arrangement, and availability fuels. Relationships between climate variability, climate change, and wildfire interact with other factors that influence fuels (Stephens, 2005). In the Pacific Northwest, regional land use history (including timber harvest/forest clearing, fire suppression and possibly fire exclusion through grazing) has affected the amount and structure of fuels. This is particularly evident for drier forests in

---

[6] Author: Jeremy Littell, University of Washington

19

the eastern Cascades, Blue Mountains, and northwestern U.S. Rockies in Washington, Idaho, Oregon, and western Montana, where fire suppression has resulted in increased fire return intervals (e.g., Hessl et al. 2004; Heyerdahl et al. 2002, 2008a and 2008b) and likely less so in wetter forests (e.g., maritime Pacific northwest coast of Oregon and Washington).

Despite changes in land use and the resulting effects on fuels, climatic correlates with fire area burned and the number of large fires are consistent in both the pre-settlement period and the last few decades. The impacts of increases in temperature and changes in precipitation affect fuel amount, structure, and availability via seasonal influences on water balance and effects on fuel moisture during the fire season. They also affect the length of the fire season. Syntheses of fire-climate relationships for both pre-settlement and modern records exist in several subregions of the West. Fire history studies (evidence from trees scarred by fires or age classes of trees established after fire and independently reconstructed climate) and modern fire-climate comparisons (evidence from observed fire events and observed climate occurring in the seasons leading up to and during the fire) agree on basic mechanisms, which vary with forest and region (Westerling et al. 2003; Littell et al. 2009a). In each forest type, climate may affect the availability of existing fuels to fire through drought or increased temperature (PNW: Heyerdahl et al. 2002, Hessl et al. 2004, Heyerdahl et al. 2008a; Northern Rockies: Heyerdahl et al. 2008b; Westside: Westerling et al. 2003, Littell et al. 2009a). Climate may instead influence fire through increased precipitation increases the availability of new, fine fuels through vegetation growth which then becomes available fuel in subsequent seasons or years (Swetnam and Betancourt, 1998; Littell et al. 2009a). Some forests have elements of both (Littell et al. 2009a).

The impact of climate change on forest fires has been assessed using statistical models that project area burned from climate variables (West wide: McKenzie et al. 2004, Spracklen et al. 2009, Littell, 2010; Pacific Northwest: Littell et al. 2010). The range of changes in area burned projected in these studies is from <100% increase in area burned to > 500% increase in median area burned depending on the time frame, methods, future emissions and climate scenario, and region. Dynamic vegetation models have also been used to project future fire activity, suggesting a range (from declines of -80% to increases of +500% depending on region, climate model, emissions scenario) of changes in biomass area burned based on climate projections derived from prior-generation GCMs over the West (Bachelet et al. 2001). Fire area burned in PNW forests is sensitive to climate (McKenzie et al. 2004; Littell et al. 2009a, 2010). The projected impacts of climate change on fire in the PNW are generally for increased fire area burned and biomass consumed in forests. Littell et al. (2010) used statistical climate-fire models to project future median regional (WA, OR, ID, w. MT) area burned increases from about 0.2 million ha (0.5 M ac) to 0.3 million hectares (0.8 M ac) in the 2020s, 0.5 million ha (1.1 M ac) in the 2040s, and 0.8 million ha (2.0 M ac) in the 2080s (Littell et al. 2010, average of CGCM3 and ECHAM5 GCMs for A1B emissions). The area burned is expected to increase on average (A1B and B1 ensemble mean climate) by a factor of 3.8 in forested ecosystems (Western and Eastern Cascades, Okanogan Highlands, Blue Mountains) to 1980–2006 (Littell et al. 2010). Rogers et al. (2011) used the MC1 dynamic vegetation model to project fire given climate and dynamic vegetation under three GCMs (A2 emissions, CSIRO Mk3, MIROC 3.2 med-res, and

Hadley CM3) showed large increases in area burned (76%–310%, depending on climate and fire suppression scenario) and burn severities (29%–41%) by the end of the twenty-first century compared to 1971 - 2000.

There is a comparative lack of quantitative information on likely forest fire frequency, severity and intensity responses to climate change. Fire area increases imply increases in fire frequency for any definable unit, but that timing of these is uncertain relative to the mid- and late-21st century because fire return intervals vary from less than 10 to over 500 years naturally within the region. Fire severity (proportion of overstory mortality) is potentially influenced by climate, though severity may be more sensitive to the arrangement and availability of fuels (which affect intensity) than area burned and so the climate effects in the future are less certain. Extreme weather events and self-driven weather consistent with large fires suggests greater severity could be expected in forest systems. To my knowledge, there are no peer-reviewed syntheses of climate-fire severity effects or projections of future severity as a function of climate.

The risk posed by future fire activity in a changing climate is a function of the likely impacts to human and ecological systems and there are important implications for adaptation and vulnerability. At the wildland / urban interface, changes in population combined with changes in forest density and/or area present forest conditions that are likely to experience increases in area burned and possibly fire severity greater than in the historical record. Fire risk is therefore likely to increase in a warmer climate due to the increased duration of the fire season, increased availability of fuels if temperature increases and precipitation does not increase sufficiently to offset summer water balance deficit. There are also likely to be influences on managed forests (private, federal, state), which have additional economic implications. In systems where fuels management (particularly using prescribed fire) is common, adaptation of forest fuels to current and future climate is an ongoing process, and so risk can be potentially mitigated. Finally, future fire risk may depend as much on whether extreme fire weather conditions will change as monthly to seasonal climate changes. Even if fire weather and ignitions don't change, it is likely that risk driven only by seasonal climate changes will increase, particularly in the wildland urban interface and managed forests, where fire has been historically rare or fully suppressed and climate has not been as strong an influence as in wildland fires.

Consequences:

Category	Consequences (**bold** denotes a positive consequence)
Natural Systems	• Decrease water quality from post fire runoff sedimentation & particulates • Increased erosion into tributary streams • Loss of forest, rangeland, & natural resources (e.g. older/complex forests=>younger forests; shrubland/sagebrush steppe=>grassland) • Disturbance of aquatic habitat & productivity • Changes to seasonal hydrograph after fire disturbance (e.g.

	larger peak flows & flashier response to precipitation events) • Increased CO2 to atmosphere • Habitat degradation & loss of connectivity (i.e. isolation of population & ecotypes) • Loss of forest biodiversity • Increase opportunity for invasive/alternative species • **Fire maintains occurrence & abundance of some tree species** • **Wildfire as a tool to promote resilience of ecosystem function to environmental change by facilitating migration and other processes**
Managed Systems (e.g. Agriculture, Forestry)	• Loss of forest productivity/resources • Forest mortality (e.g. high probability in East Cascades and Blue Mountain ecoregions in next 20-40 years, Littell et al. 2009b) • Large scale disturbance of forest systems (e.g. conversion, extirpation, extinction, facilitation, loss of structure)
Built Environment	• Damages to homes & property & other encroaching built infrastructure (urban/wildland interface) • Damage to electricity transmission lines causing widespread power outages
Economy	• Potential loss of commercial timber & revenues • High fire management costs • Major financial cost for private land/cabin owners
Human Society	• Increase in poor air quality days (especially East side) • Loss of livelihood & cultural resources (e.g. timber dependent communities; native American communities in sagebrush steppe) • Cultural changes in forest landscape (e.g. private forest lands are no longer profitable in forestry, but become converted lands) - *high* • Decreased drinking water quality from particulates & retardants in surface water supply • Increase in injuries/fatalities & home displacement from fire

Summary of Survey Rating Questions:

Human Population Affected: Half of the respondents indicated that few people would be affected by an increase in wildfires. A common reason reported was that the forested areas of high wildfire risk in the Northwest are not highly or densely populated. Though the people who do live in areas at risk to fire are certainly threatened, including development. The risk of smoke inhalation and potential asthma cases has a farther-reaching effect on the population.

Human Mortality: About half of the respondents said there would likely be an increase in human mortality due to increased wildfires with the most common cause due to smoke inhalation or air quality issues. One comment indicated that there is some evidence that there is already an increase in mortality from wildfires, and the trend may generally increase over time and may depend on adaptation and preparedness techniques. The other roughly half of the respondents indicated uncertainty in whether human mortality would increase or decrease.

Human Health Quality: About three-fourths of respondents indicated there would be a reduction in general health quality due to wildfires, but not an extreme reduction. The most common cause is respiratory issues to smoke exposure and reduced air quality, especially for those with existing respiratory problems. About one-fifth chose a response that did not indicate a sign of change regarding consequences of wildfires to human health quality, but a few people mentioned respiratory issues from smoke inhalation as an potential consequence.

Biodiversity: Just over 60% of respondents indicated that an increase in wildfire would be unfavorable for biodiversity and just over 15% indicating a severe reduction in biodiversity. Forest vegetation mortality may lead to a loss of habitat for small birds and animals and niche species. Species composition will likely change after wildfires and there may be a shift to generalist or invasive species, through repopulation, at the expense of specialist or native species. However, traditional use of fire may have beneficial effects on local ecosystems. About one-fifth of respondents indicated they didn't know.

Geographic area affected: Just less than two-thirds (64.1%) of respondents indicated that a moderate portion of the region would be affected by increasing wildfires. About 18% of respondents indicated that most of the region would be affected. Roughly half of the region is forest cover (Littell et al. 2009b; ODF; USFS) with some areas highly disrupted already like the Blue Mountains, Cascades, and grass areas. Areas with low moisture, disease/infestations, and wind are more susceptible to wildfire, like Eastern Washington.

Built Environment: About half of the respondents indicated a moderate disruption of public services and infrastructure due to increasing wildfires. About 22% indicated little disruption, and about 14% indicated major disruption. Within the comments, two main themes of disruption to the built environment were: direct disruption to infrastructure and communities on the urban/forest interface and indirect disruption to water management facilities including cost increase of water supply filtration. A few of those who reported little disruption also indicated impacts on the built environment at the urban/forest interface, but that it constituted a small portion of the built environment.

Economy: About two-thirds of respondents indicated either medium or high economic loss due to increasing wildfires. About one quarter of respondents indicated they didn't know. It was noted that it is difficult to estimate economic losses because it depends on the timescale, geographic vulnerabilities, size of economy, forest productivity changes, infrastructure damage, and fire management. However, it is generally corroborated that there would be economic losses due to increasing wildfires, especially to the timber industry and built environment. A review of historical economic impacts of wildfire is needed.

The breakout group identified two broad categories of consequences: ecosystem and biophysical impacts and impacts to human health, well-being, and infrastructure. While the group chose not to rank consequences, most of the consequences would either be characterized as high consequence and high likelihood, or high consequence with low or uncertain likelihood in the short term, but expected eventually. The degree of consequences varies across sub-regional climates, vegetation or ecosystem type, and fire regime. When considering the consequences of increases in wildfire, geographic heterogeneity and severity of consequences need to be taken into account. Disaggregating consequences into East and West parts of the region might facilitate a better assessment of the magnitude, as consequences could be larger and less predictable in the Cascades and Coast Range. It may also be helpful to distinguish between acute risk (e.g. house burning) and chronic consequence (e.g. smoke in rural communities).

There was also a noteworthy discussion of the cultural impacts on timber communities, particularly on the west side of the region, and shrub steppe communities on the east side of the region. A high consequence impacting forest landowners, workers, timber-dependent communities, and tribes, could be the change in development pressure dynamic for private forests, that is, if economic consequences of increasing wildfires cause lands to become unprofitable for timber production and are converted. The shrub steppe system may be more likely than forest systems to experience fire-driven step changes that could occur quickly and are due to a combination of factors including invasive grasses and fire feedbacks. The potential consequence of conversion of steppe to grassland is a loss in biodiversity of flora and fauna, and could adversely impact the Native American communities living in the sagebrush steppe regions.

References:

Bachelet, D., R. P. Neilson, J. M. Lenihan, and R. J. Drapek, 2001. Climate change effects on vegetation distribution and carbon budget in the United States. Ecosystems 4:164 - 185

Elsner, M. M., L. Cuo, N. Voisin, et al., 2010. Implications of 21st century climate change for the hydrology of Washington State. Climatic Change 102:225-260.

Hardy, C.C. 2005. Wildland fire hazard and risk: problems, definitions, and context. Forest Ecology and Management. 211: 73–82.

Henderson, J. A., D. H. Peter, R. D. Lesher, and D. C. Shaw. 1989. Forested Plant Associations of the Olympic National Forest. USDA Forest Service, Pacific Northwest Region. R6-ECOL-TP 001-88. 502 p.

Hessl, A. E., D. McKenzie, and R. Schellhaas, 2004. Drought and Pacific Decadal Oscillation linked to fire occurrence in the Inland Pacific Northwest. Ecological Applications 14:425-442.

Heyerdahl, E. K., L. B. Brubaker, and J. K. Agee, 2002. Annual and decadal climate forcing of historical regimes in the interior Pacific Northwest , USA. The Holocene 12:597-604.

Heyerdahl, E. K., D. McKenzie, and L. D. Daniels, et al., 2008a. Climate drivers of regionally synchronous fires in the inland Northwest (1651–1900). International Journal of Wildland Fire 17:40-49.

Heyerdahl, E. K., P. Morgan, J. P. Riser, 2008b. Multi-season climate synchronized historical fires in dry forests (1650-1900), northern Rockies, U.S.A. Ecology 89:705-16.

Littell, J. S., D. McKenzie, D. L. Peterson, and A. L. Westerling, 2009a. Climate and wildfire area burned in western U . S . ecoprovinces, 1916-2003. Ecological Applications 19:1003-1021.

Littell, J. S., E. E. Oneil, D. McKenzie, J. A. Hicke, J. Lutz, R. A. Norheim, M. M. Elsner, 2009b. Forest ecosystems, disturbance, and climatic change in Washington State, USA. The Washington Climate Change Impacts Assessment, M. McGuire Elsner, J. Littell, and L Whitely Binder (eds). Center for Science in the Earth System, Joint Institute for the Study of the Atmosphere and Oceans, University of Washington, Seattle, Washington. Available at:
http://www.cses.washington.edu/db/pdf/wacciareport681.pdf

Littell, J. S. 2010. Figure 5.8, Chapter 5, "Impacts in the next few decades and the next century, Fire and climate in "Climate Stabilization Targets: Emissions, Concentrations, and Impacts over Decades to Millennia (2011, Board of Atmospheric Sciences and Climate, National Research Council)". http://www.nap.edu/catalog.php?record_id=12877 page 178 and 180.

Littell, J. S., E. E. Oneil, D. McKenzie, et al., 2010. Forest ecosystems, disturbance, and climatic change in Washington State, USA. Climatic Change 102:129-158.

Littell, J. S. and R. Gwozdz, 2011. Climatic Water Balance and Regional Fire Years in the Pacific

Northwest, USA: Linking Regional Climate and Fire at Landscape Scales. Chapter 5 in McKenzie, D., C. M. Miller, and D. A. Falk, Eds. The Landscape Ecology of Fire, Ecological Studies 213, DOI 10.1007/978-94-007-0301-8_5, © Springer Science+Business Media B.V. 2011

Marlon, J. R., P. J. Bartlein, C. Carcaillet, et al., 2008. Climate and human influences on global biomass burning over the past two millennia. Nature Geoscience 1:697-702.

McKenzie, D., Z. Gedalof, D. L. Peterson, Mote, P., 2004. Climatic change, wildfire, and conservation. Conservation Biology 18:890-902.

Oregon Department of Forestry (ODF):
http://www.oregon.gov/ODF/PUBS/docs/Forest_Facts/FFForestryFactsFigures.pdf. Accessed: February 24, 2012

Peterson and Vose, et al. National Climate Assessment Forest Sector Technical Input. In preparation.

Power, M. J., J. Marlon, N. Ortiz, et al., 2007. Changes in fire regimes since the Last Glacial Maximum: an assessment based on a global synthesis and analysis of charcoal data. Climate Dynamics 30:887-907.

Rogers, B. M., R. P. Neilson, R. Drapek, et al., 2011. Impacts of climate change on fire regimes and carbon stocks of the U.S. Pacific Northwest. Journal of Geophysical Research 116:1-13.

Spracklen, D. V., L. J. Mickley, J. A. Logan, et al., 2009. Impacts of climate change from 2000 to 2050 on wildfire activity and carbonaceous aerosol concentrations in the western United States. Journal of Geophysical Research 114:1-47.

Stephens, S. L., 2005. Forest fire causes and extent on United States Forest Service lands. International Journal of Wildland Fire 14:213-222.

U.S. Forest Service (USFS): http://www.fs.fed.us/rm/ogden/overviews/Idaho/OV_Idaho.htm. Accessed: February 24, 2012.

Westerling, A. L., A. Gershunov, T. J. Brown, et al., 2003. Climate and wildfire in the western United States. Bulletin of the American Meteorological Society 84:595-604.

Westerling, A. L., H. G. Hidalgo, D. R. Cayan, and T. W. Swetnam, 2006. Warming and earlier spring increase western U.S. forest wildfire activity. Science 313:940-3.

Likelihood:[7]

The likelihood of chemistry changes in association with ocean acidification is high because it is already occurring and documented for Pacific Northwest waters. It is well established that human activities are changing the chemistry of the ocean (Feely et al. 2008; Doney et al. 2009; NRC, 2010) by causing atmospheric CO_2 levels to climb higher than at any time in at least the past 650,000 years. Current estimates are that about 30% of the anthropogenically-derived CO_2 released to the atmosphere over the past 250 years is now dissolved in the ocean (Canadell et al. 2007; Feely et al. 2010). Once dissolved in the ocean, CO_2 causes the pH and carbonate saturation state of seawater to decline, rendering ocean water corrosive with respect to the calcite and aragonite shells and skeletons of marine organisms. These changes, commonly referred to as ocean acidification, are larger and are occurring faster than at any time in the past hundreds of thousands to millions of years. The persistence of contemporary marine ecosystems is threatened, as is the persistence of shellfish aquaculture and the human communities that depend on them.

A combination of factors renders the Pacific coast of North America especially vulnerable to acidified "corrosive" water events. Deep waters of the Pacific are the oldest in the world's oceans and, to a degree, are naturally corrosive from accumulated natural respiration processes and oxidation of organic matter. Anthropogenic additions of CO_2 further reduce the pH and carbonate saturation state of Pacific coast waters to levels that can challenge calcification in shelled organisms (Feely et al. 2008, 2010). This is exacerbated when seasonal upwelling transports these corrosive water up onto the continental shelf of the Pacific Northwest, where in some places, they reach all the way to the surface (Feely et al. 2008; Hauri et al. 2009). In coastal estuaries, inputs of nutrients and organic matter reduce pH and carbonate saturation state even further such that conditions within Puget Sound were more corrosive waters exist than observed off the coast (Feely et al. 2010). *Consequently, natural processes, anthropogenic additions of CO_2, and additions of nutrients and organic matter to estuaries and nearshore areas combine to intensify ocean acidification in Pacific Northwest coastal estuaries.* Rykaczewski and Dunne (2010) note that the Pacific coast between British Columbia and northern California is particularly productive and retentive of organic material when compared with other regions of the California Current System, and conclude that this area will be impacted earlier and more intensely than other North American coastal regions by ocean acidification.

The same Pacific Northwest coastal estuaries that are threatened by ocean acidification are the source of highly valued shellfish fisheries. In Washington alone, shellfish growers in 2005 produced approximately 90 million pounds of shellfish with an estimated value of $97 million (PCSGA, 2010). Shellfish aquaculture provides an important source of jobs in Washington, and revenues directly benefit state and local economies. Loss of shellfish aquaculture from the Pacific Northwest would impose substantial social and economic costs.

---

[7] Authors: Jan Newton, University of Washington; Kathie Dello, Oregon State University

Bivalves are known to be highly vulnerable to reductions in pH and carbonate saturation state (Green et al. 2004; Gazeau et al. 2007; Talmadge and Gobler, 2009; Hettinger et al. 2010). Projections based on future scenarios indicate that mussel and oyster calcification rates could decline by as much as 25% and 10%, respectively, by the end of this century (Gazeau et al. 2007; Ries et al. 2009). Larval and juvenile stages are particularly sensitive to corrosive water conditions. Growth rates in larval and juvenile stages of the Olympia oyster were up to 41% slower under high CO_2 conditions (970 ppm) compared with growth in controls (Hettinger et al. 2010). Slower growth rates persisted among treated oysters even after they had been restored to control conditions that matched present-day levels of CO_2 in seawater, suggesting the existence of legacy or carry-over effects from larval to adult stages.

The likelihood of rising ocean temperatures is high because it is already occurring and documented for Pacific Northwest waters, but there is considerable spatial and temporal variability to be expected in the signal. Ocean heat content and average sea surface temperatures (SSTs) have been increasing on a global-ocean scale (Bindoff et al. 2007; Trenberth et al. 2007). Models predict PNW coastal SST to increase by 1.2 °C by the 2040s (Mote and Salathé, 2010). Locally, the coastal upwelling / downwelling cycle leads to strong variation in temperature annually; the average sea surface temperature in near coastal environments varies by about 8°C seasonally (Mote and Salathé, 2010). Changes in upwelling could cause more or less dramatic temperature shifts, depending on what wind shifts occur and the timing of such. SST will be highly influenced by several weather-related factors, such as wind and air temperature, as well as ocean-related factors, such as upwelling, mixing, stratification, as well as factors associated with geography such as proximity to rivers and bathymetric features that cause turbulent mixing. Thus strong spatial variation in the PNW can be expected.

Moore et al. 2008 investigated the influence of climate on Puget Sound oceanographic properties on seasonal to interannual timescales using continuous profile data at 16 stations from 1993 to 2002 and records of sea surface temperature (SST) and sea surface salinity (SSS) from 1951 to 2002. Variability in the leading pattern of Puget Sound water temperature and salinity profiles was well correlated with local surface air temperatures and freshwater inflows to Puget Sound from major river basins, respectively. SST and SSS anomalies were informative proxies for the leading patterns of variations in Puget Sound temperature and salinity profiles. They found that SST and SSS anomalies also have significant correlations with Aleutian Low, El Niño-Southern Oscillation, and Pacific Decadal Oscillation variations in winter that can persist for up to three seasons or reemerge the following year. However, correlations with large-scale climate variations were weaker compared to those with local environmental forcing parameters. As climate change will affect both local and large-scale forcings, effects on SST and SSS may be complex.

Consequences:

Category	Consequences (**bold** denotes a positive consequence)

Natural Systems	• Disruption to food chain, especially loss of important species at the base and compounding effects upwards *(high)* • Species affected negatively: Salmon (health & mortality if warmer ocean temps => increase hypoxic events), Lamprey, Shellfish (establishment & growth), calcifying organisms, migratory birds • Increase algal blooms affecting water quality & migratory birds • Disrupt ocean organisms & ecosystems & biodiversity (see Tillmann and Siemann, 2011b) *(medium)* • High risk to marine life in upwelling zones & estuaries
Managed Systems (e.g. Agriculture, Forestry)	• Negative impacts on oyster recruitment in hatcheries (already seen in last 5 years, see Grossman article; Barton et al. in press) • Reduction in shellfish development impacting the industry (lack of natural recruitment (Dumbauld et al. 2010))*(high)* • Loss numbers of returning salmon • Affects estuary aquaculture (i.e. Coquille tribe has a cranberry bog) *(medium)*
Built Environment	• Decrease salmon health affects Columbia River management
Economy	• Disruption to seafood industry (i.e. harvest loss) especially shellfish, fisheries, salmon *(high)* • Affects fishing tourism *(high)* & beach tourism *(low)*
Human Society	• Impacts on fisheries dependent communities *(high)* • Human health effects *(medium)* (e.g. from increase harmful algal blooms)

Summary of Survey Rating Questions:

Human Population Affected: Of the two most common response choices, respondents were divided between few people affected (36.8%) and most of the population affected (28.9%). However, comments indicated that most coastal communities and tribal populations that depend on coastal marine resources will likely be affected through impacts to the economy from ocean harvest or a change in diet from decreased harvest of fisheries and shellfish and a potential collapse of the marine food web over the long term.

Human Mortality: About 80% of respondents chose a response that didn't indicate a sign of change in human mortality. Changes in ocean temperature and chemistry are likely not a factor contribution to increased human mortality. Just less than one-fifth of respondents indicated an increase in mortality. There is a slight potential for increased illnesses and deaths associated with harmful algal blooms if they were to increase in frequency and severity in the future, but that is unlikely.

Human Health Quality: About two-thirds of respondents chose a response that did not indicate a sign of change for human health quality. Many felt that changes in health quality would not likely be a factor. About one-third indicated that a reduction in general health quality is to be expected. Two factors mentioned in terms of contributing to potential decrease in human health quality are loss of protein in diets from decreased ocean harvests and illnesses related to harmful algal blooms and consumption of contamination of shellfish.

Biodiversity: The majority felt biodiversity would be negatively affected by ocean temperature and chemistry changes. 47.4% of respondents indicated a severe reduction in biodiversity, while 28.9% indicated unfavorable for biodiversity. All the comments indicated a negative affect or loss to marine organisms and habitats including slow-moving species from hypoxic areas, shell-forming organisms, effects of a cascading food web, and effects from ocean acidification occurring too quickly for species to adapt. There is little adaptive capacity on the part of humans to protect marine organisms, but individual species may or may not have adaptive capacity.

Geographic area affected: Over four-fifths of respondents felt that either a moderate portion (as indicated by 50% of respondents) or most of the region would be affected by ocean temperature and chemistry changes. Comments indicated that response choice depends on the definition of 'region', whether it be the entire Northwest or just the coastal Northwest. In any case, respondents felt the entire Northwest coastal area including estuaries and the Puget Sound would be affected, but with regional spatial and temporal heterogeneity.

Built Environment: Over half (56.8%) of respondents felt that there would be little disruption in public services and infrastructure due to ocean temperature and chemistry changes. Ocean acidification could affect coastal and offshore structures such as pipelines, cooling intakes or other metal structures, though respondents were uncertain of the extent of this possibility. There could be indirect effects on coastal infrastructure due to sea level rise caused by ocean temperature increase and thermal expansion.

Economy: The respondents who provided rationale with their ratings converged on two main industries for potential large economic losses: shellfish industry (negative impacts already being seen) and coastal fisheries, including salmon. There is uncertainty in the extent of economic loss. The coastal tourism industry is also suggested as a possible contributor to economic loss.

Case Study Examples:

A workshop participant highlighted the Coquille Tribe cranberry farmers of Coos Bay, Oregon as an example of the effects of ocean acidification.

Another compelling example with strong economic impacts is the effect of ocean acidification on the survival rates of shellfish at the Whiskey Creek and Taylor shellfish hatcheries (see Grossman article; Barton et al., in press) as well as the lack of economically viable natural recruitment of oysters in Willapa Bay since 2005 (Dumbauld et al. 2011). The economic losses have been estimated at roughly $46 million dollars in this region alone (source: Sustainable Fisheries Partnership).

Identified Gaps in Knowledge:

Major uncertainty still exists in terms of how coastal winds, and hence upwelling, will change with climate change and regarding a possible link between warmer ocean temperatures and hypoxic events. Lack of timeseries data and adequate spatial coverage make risk assessments challenging. Experiments with native organisms and local/expected conditions are only starting to gain traction so there are fewer results to draw from.

Special Considerations:

Pacific waters are susceptible to being corrosive due to the long residence time of the deep Pacific waters, seasonal upwelling that brings these deep waters in contact with coastal areas, and potential for further carbon and nutrient loading from both natural and anthropogenic sources. These factors combine to make the Pacific Northwest one of the first regions that will likely see the effects of ocean acidification. Another factor contributing to changing ocean temperature and chemistry is changes in river flow and timing from rivers like the Columbia and Fraser.

References:

Bindoff NL, Willebrand J, Artale, Cazenave VA, Gregory J, Gulev S, Hanawa K, Le Que´re´ C, Levitus S, Nojiri Y, Shum CK, Talley LD, Unnikrishnan A (2007) Observations: oceanic climate change and sea level. In: Solomon S, Qin D, Manning M, Chen Z, Marquis M, Averyt KB, Tignor M, Miller HL (eds) Climate change 2007: the physical science basis. Contribution of Working Group I to the Fourth Assessment Report of the Intergovernmental Panel on Climate Change. Cambridge University Press, Cambridge

Barton, A., B. Hales, G.G. Waldbusser, C. Langdon, and R.A. Feely. In press. The Pacific oyster, Crassostrea gigas, shows negative correlation to naturally elevated carbon dioxide levels: Implications for near-term ocean acidification effects. *Limnology and Oceanography.*

Canadell, J.G., C. Le Quere, M.R. Raupach, C.B. Field, E.T. Buitenhuis, P. Ciais, T.J. Conway, N.P. Gillett, R.A. Houghton, and G. Marland. 2007. Contributions to accelerating atmospheric CO_2 growth from economic activity, carbon intensity, and efficiency of natural sinks. Proceedings of the National Academy of Sciences of the United States of America, 104, 18866-18870.

Doney, S.C., V.J. Fabry, R.A. Feely, and J.A. Kleypas. 2009. Ocean acidification: the other CO_2 problem. Annual Review of Marine Science, 1: 169-192. DOI: 10.1146/annurev.marine.010908.163834.

Dumbauld, B.R., B.E. Kauffman, A.C. Trimble, and J.L. Ruesink. 2011. The Willapa Bay oyster reserves in Washington State: Fishery collapse, creating a sustainable replacement, and the potential for habitat conservation and restoration. *Journal of Shellfish Research,* 30: 71-83.

Feely, R.A., C.L. Sabine, J.M. Hernandez-Ayon, D. Ianson, and B. Hales. 2008. Evidence for upwelling of corrosive "acidified" water onto the continental shelf. Science, 320, 1490-1492.

Feely R A., S.R. Alin, J.A. Newton, C.L. Sabine, M. Warner, A. Devol, C. Krembs, and C. Maloy. 2010. The combined effects of ocean acidification, mixing, and respiration on pH and carbonate saturation in an urbanized estuary. Estuarine, Coastal and Shelf Science, 88 (2010) 442-449.

Gazeau, F., C. Quiblier, J.M. Jansen, J.P. Gattuso, J.J. Middelburg, and C.H.R. Heip, 2007. Impact of elevated CO_2 on shellfish calcification. Geophysical Research Letters, 34.

Green, M.A., M.E. Jones, C.L. Boudreau, R.L. Moore, and B.A. Westman. 2004. Dissolution mortality of juvenile bivalves in coastal marine deposits. Limnology and Oceanography, 49, 727-734.

Grossman, Elizabeth, 2011: Northwest oyster die-offs show ocean acidification has arrived. Yale Environment 360, a publication of the Yale School of Forestry & Environmental Studies. (http://e360.yale.edu/feature/northwest_oyster_die-offs_show_ocean_acidification_has_arrived/2466/)

Hauri, C., N. Gruber, G.-K. Plattner, S. Alin, R.A. Feely, B. Hales, and P.A. Wheeler. 2009. Ocean acidification in the California Current System. *Oceanography* 22(4):60–71, http://dx.doi.org/10.5670/oceanog.2009.97.

Hettinger, A., E. Sanford, B. Gaylord, T. M. Hill, A. D. Russell, M. Forsch, H.N. Page, and K. Sato. 2010. Ocean acidification reduces larval and juvenile growth in the Olympia oyster (*Ostrea lurida*), Eos Trans. AGU, 91(26), Ocean Sci. Meet. Suppl., Abstract BO51A-06.

Mote, P. and E.P. Salathé, 2010: Future climate in the Pacific Northwest. *Climatic Change* 102(1-2): 29-50, doi: 10.1007/s10584-010-9848-z.

Moore, S. K., N. J. Mantua, J. P. Kellogg, and J. A. Newton. 2008. Local and large-scale climate forcing of Puget Sound oceanographic properties on seasonal to interdecadal timescales. *Limnology and Oceanography,* 53: 1746-1758.

National Research Council (NRC). 2010. Ocean Acidification: A National Strategy to Meet the Challenges of a Changing Ocean. Committee on the Development of an Integrated Science Strategy for Ocean Acidification Monitoring, Research, and Impacts Assessment. National Research Council. http://www.nap.edu/catalog/12904.html

Pacific Coast Shellfish Growers Association 2010 (http://www.pcsga.org/pub/farming/production_stats.shtm)

Ries, J.B., A.L. Cohen, and D.C. McCorkle. 2009. Marine calcifiers exhibit mixed responses to CO_2-induced ocean acidification. Geology, 37, 1131-1134.

Rykaczewski, RR and JP Dunne. 2010. Enhanced nutrient supply to the California Current Ecosystem with global warming and increased stratification in an earth system model. *Geophys. Res. Lett.* **37**:L21606, doi:10.1029/2010GL045019.

Talmage S.C. and C.J. Gobler. 2009. The effects of elevated carbon dioxide concentrations on the metamorphosis, size and survival of larval hard clams (*Mercenaria mercenaria*), bay scallops (*Argopecten irradians*), and Eastern oysters (*Crassostrea virginica*). Limnol. Oceanogr., 54: 2072-80.

Tillmann, P. and D. Siemann, 2011b. Climate Change Effects and Adaptation Approaches in Marine and Coastal Ecosystems of the North Pacific Landscape Conservation Cooperative Region: A Compilation of Scientific Literature. Phase 1 Draft Final Report. National Wildlife Federation – Pacific Region, Seattle, WA. August 2011.

Trenberth, K. E., et al. (2007), Observations: Surface and atmospheric climate change, in Climate Change 2007: The Physical Science Basis. Contribution of Working Group I to the Fourth Assessment Report of the Intergovernmental Panel on Climate Change, edited by S. Solomon et al., pp. 235–336, Cambridge Univ. Press, Cambridge, U. K.

What is the likelihood of increased coastal erosion and risk of inundation from increasing sea levels, increasing wave heights, and storm surges occurring mid-21st century relative to the late 20th century? – HIGH

Earth's changing climate is expected to have significant physical impacts along the coast and estuarine shorelands of the Northwest, ranging from increased erosion and inundation of low lying areas to wetland loss. The environmental changes associated with climate change include rising sea levels, possibly the increased occurrences of severe storms, and rising air and water temperatures. The combination of these processes and their climate controls are important to beach and property erosion and flood probabilities, with the expectation of significant changes projected for the 21st century.

Coastal change and flood hazards along the Northwest coast are caused by a number of ocean processes, each of which has significant climate controls such that the severity and frequency of the hazards in the future can be expected to increase. Not only will global sea level almost certainly increase throughout the 21st century, by perhaps a meter or more, there is good reason to believe that the rate of sea level rise will itself increase (Rahmstorf, 2007; Rahmstorf, 2010). Evaluating the consequences of intensified and more frequent hazards is complicated by the tectonic setting of the Northwest, with there being significantly different rates of land uplift along the coast. Taken together, the variable rate of uplift plus the present-day rate of sea level rise, some stretches of the coast are presently submerging as the sea-level rise is greater than the tectonic uplift, whereas other areas are emerging where the reverse is true (Komar et al. 2011). The prospects are that with accelerated rates of sea-level rise the entire coast will eventually be submerging and experience significantly greater erosion and flood impacts than at present day.

Allan et al. (2011) analyzed the Yaquina Bay storm surge record and found no increases in surge levels and frequencies since the late 1960s. Several recent studies have documented increasing wave heights and in particular increasing extreme wave heights in the region from analyses of buoy data (e.g., Allan and Komar, 2006; Menendez et al. 2008, Ruggiero et al. 2010). However these results have recently been called into question after careful analyses of modifications of the wave measurement hardware as well as the analysis procedures since the start of the observations have demonstrated inhomogeneities in the records (Gemmrich et al. 2011). On the other hand, the satellite altimetry record from 1985 – 2008 reveals increases in extreme wave heights (99th percentiles) and wind speeds in the region (Young et al. 2011).

The periodic occurrence of major El Niño events in the future will compound the impacts of increasing sea levels, resulting in severe episodes of coastal erosion and flooding, as

---

[8] Author: Peter Ruggiero, Oregon State University. Note that much of the text is pulled directly from the Oregon Climate Assessment Report chapter on *Impacts of climate change on Oregon's coasts and estuaries*, of which Ruggiero is the first author.

experienced during the El Niño winters of 1982-83 and 1997-98. At present it is not known whether El Niño intensity and frequency will increase under a changing climate. With these multiple processes and their climate controls having important roles in causing erosion and flooding along the coast, it is challenging to collectively analyze them with the goal of providing meaningful assessments of future coastal hazards during the next several decades.

Coastal infrastructure will come under increased risk to damage and inundation under a changing climate with impacted sectors including transportation and navigation, coastal engineering structures (seawalls, riprap, jetties etc.) and flood control and prevention structures, water supply and waste/storm water systems, and recreation, travel and hospitality. Situated just slightly above current sea level, the major ports of Seattle and Tacoma in the Puget Sound of Washington are making plans to raise heights of some port infrastructure in response to sea level rise (CIG, 2009).

Unfortunately, significant knowledge gaps remain, impairing our ability to accurately assess the impacts of climate change along our coast and estuarine shorelands. The NCA Sea Level Change Scenarios team has medium high confidence in a 0.2 to 2 meter rise in global mean sea level by 2100. At present we do not conclusively understand the climate controls on changing patterns of storminess and wave heights and therefore have a very limited ability to project future trends in coastal storm impacts. The magnitude and frequency of major El Niño events has significant implications for the Northwest, however, at this time we are unable to assess whether or not these will increase in the future due to climate change. Further, the long-term time-series data necessary to definitively identify perturbations of estuarine communities that can be attributed to anthropogenic climate change are lacking and therefore our understanding of anticipated shifts remain largely speculative.

Consequences:

Category	Consequences (**bold** denotes a positive consequence)
Natural Systems	• Estuarine impacts (increased salinity, sediment transport) • Soil salinization • Beach, bluff, dune loss from erosion • Compromised coastal habitats (intertidal, wetlands – important nursery habitat, tidal marsh, etc.) • Threatened/Endangered Species (snowy plover) • Loss of species from SLR (seabirds, shorebirds, waterbirds, etc.) • Alter food web • Mobilization of toxics affects ecosystems (especially industrial areas around Puget Sound, Columbia River, & Astoria) • **Increase in shallow water habitat**
Managed Systems (e.g. Agriculture, Forestry)	• Altered location of aquaculture • Increase flooding in low lying agricultural areas and tribal land (e.g. Skagit Valley farming; Coquille cranberry farming)

Built Environment	• Property & Infrastructure loss or damage (roads, railroads, waste water, houses, businesses, docks, seawall, dikes, levees, harbor reconfiguration, ports, private property & land, bridges) • Transportation – (Increased maintenance cost of roads, abandoning roads influencing accessibility) • Protect private structures – (rip rap, businesses that depend on proximity to shore, relocation
Economy	• Decreased property/real estate value • Altered recreational value (tourism) • Altered commercial fisheries value • Shellfish aquaculture industry • Shipping - bar closures • **Extreme wave heights could be beneficial for wave energy production** • **Increase in storm-watching tourism**
Human Society	• Water treatment plants/water quality implications from storm surge inundation • Increase risk of storm-related injuries/death from higher waves • Saltwater intrusion in coastal aquifers • Algal blooms from increased storminess causing toxicity and water closures affecting human health (e.g. shellfish) • Mobilization of toxics in industrial areas (especially Puget Sound, Columbia River, & Astoria) • Tribal losses of land, homes, cultural resources (tribal trust lands with no compensation) • Displacement of communities • Decreased accessibility from abandoned roads • May lose public access to beaches (shoreline management act)

Summary of Survey Rating Questions:

Human Population Affected: Just fewer than 50% of respondents indicated that few people would be affected by sea level rise and coastal hazards; about 20% indicated about half of the population would be affected; and 13.2% indicated that most of the population would be affected. While over half of the US population resides in coastal zones (NOAA Ocean), a smaller percentage of the Northwest's population actually lives in coastal zones. In Oregon, about 6.5% of the population lives within 50 miles of the coast (OCMP), but a much greater number of people live in counties along the Washington coast (NOAA Washington), which are seeing a large coastal population growth rate. There are low population densities within coastal communities, especially in Oregon (29 people per square mile, OCMP), excluding population centers like the Seattle area. Those at great risk include people who live near the shoreline on vulnerable bluffs for example, and tribal communities whose reservations are located in near-shore coastal zones like tribes of the Puget Sound and Olympic Peninsula. While direct consequences of sea level rise and other coastal hazards may affect relatively few people, indirect consequences have the potential to affect a much larger portion of the population. These indirect consequences include economic effects of coastal property and infrastructure damages, effects from altered food webs, shipping disruptions, inland migration, and depends on the locations and degree of vulnerability of water facilities, roads and irrigation agriculture.

Human Mortality: About 42% of respondents indicated an increase in human mortality with comments indicating direct effects like storm events, storm surges, and flooding to people caught off guard like extreme storm-watchers. However, these changes won't happen overnight and behavioral adaptation should mitigate the risk of human loss of life. About 24% of respondents indicated that the sign was uncertain.

Human Health Quality: There could be some reduced health quality (as indicated by 27% of respondents) from losing homes and people being displaced and limiting health access, or from water quality and disease transmission or other health consequences from increased storm water and wastewater backups. However, it is difficult to determine where health impacts, especially indirect ones, could occur and so the sign of impacts to health quality of those people affected is uncertain (62.1% of respondents chose a response that did not indicate a sign of change). A reduction in public safety may be more relevant than a reduction in public health.

Biodiversity: Just over half (51.4%) of respondents indicated sea level rise and coastal hazards would be unfavorable for biodiversity. According to comments, conditions unfavorable for biodiversity could result from the disruption or loss of coastal habitats and ecosystems due to erosion or fixed infrastructure near shorelines. For example, the high biodiversity of near-shore rocky intertidal areas may suffer due to the inability of species to migrate upwards into higher coastline. Saltwater intrusion is a problem for native vegetation and irrigated agriculture. The magnitude of these consequences depends partly on the local coastal land elevation and tectonic uplift rates, which could determine vulnerable areas. However, impacts on biodiversity still remains unclear (as indicated by 24.3%), especially as there will likely be winners and losers, and much future research is needed. It was noted that biodiversity is also not the only metric to measure consequences to the natural environment.

Geographic area affected: As indicated by respondent comments, the geographic extent of areas affected by sea level rise and other coastal hazards is restricted to coastal shorelines, especially those with low near-shore gradient, the low-lying, and tidally affected areas. The Puget Sound is a good example. Inland areas affected would be low-lying estuaries like the Tillamook watershed. The extent of the area affected depends on the magnitude of sea level rise that occurs globally, and relative sea level rise locally. About 60% of respondents indicated a moderate portion of the region would be affected, about 18% and 13% indicated few isolated areas and most of the region, respectively.

Built Environment: Over half of respondents indicated a moderate disruption and 36.8% indicated a major disruption in public services and infrastructure. There is a lot of coastal infrastructure at risk of damage of degradation from sea level rise, storm surges, and erosion including: roads, buildings, utilities, ports, rails, sea walls, bridges, homes, waste water treatment plants, communities. The disruption will likely be limited to coastal regions only, with little inland disruption. The disruption will depend on the frequency of events.

Economy: Respectively, 56.8%, 16.2%, and 8.1% of respondents indicated a high, medium, and low economic loss due to sea level rise and coastal hazards. There are many valuable coastal assets including infrastructure and near-shore habitats that could be vulnerable to damages from sea level rise and other coastal hazards, but may depend on the rate of sea level rise. One respondent noted that there may already be several million dollars loss per year from storm surge damage, but no studies are available for future estimates of economic loss. Mitigation and adaptation costs could be high. Other economic disruptions include high damage repair costs, business interruptions, and relocating coastal communities and facilities.

Case Study Examples:

Infrastructure disruption could be large, not only for coastal cities like Seattle, but also for smaller coastal communities like the city of Stanwood, Washington on the Puget Sound. Large areas of Stanwood might be inundated by saltwater if sea level rose only 2 feet above mean high tide (WSDOT vulnerability map). Low-lying agricultural areas, deltas, and tribal lands are also vulnerable to increased flooding and inundation. Two examples are Skagit Valley, Washington and the Coquille Tribe cranberry farms in southern Oregon.

Identified Gaps in Knowledge:

There is little information on sediment dynamics and how that would contribute to consequences of sea level rise and increasing coastal hazards. There is also high uncertainty and variability in the ecological response to such these climate changes.

References:

Allan, J.C., P.D. Komar, and P. Ruggiero, 2011. Storm surge magnitudes and frequency on the central Oregon coast. Proceedings of the 2011 Solutions to Coastal Disasters Conference, pp. 53-64.

Allan J.C. and P.D. Komar, 2006. Climate controls on U.S. West Coast erosion processes. Journal of Coastal Research, 22, 511-529.

Climate Impacts Group (CIG), 2009. The Washington Climate Change Impacts Assessment, Executive Summary. M. McGuire Elsner, J. Littell, and L Whitely Binder (eds). Center for Science in the Earth System, Joint Institute for the Study of the Atmosphere and Oceans, University of Washington, Seattle, Washington. Available at: http://www.cses.washington.edu/db/pdf/wacciareport681.pdf

Gemmrich, J., B. Thomas, and R. Bouchard, 2011. Observational changes and trends in northeast Pacific wave records. Geophysical Research Letters, 38 (22), L22601.

Komar, P.D., J.C. Allan, and P. Ruggiero, 2011. Sea Level Variations along the U.S. Pacific Northwest Coast: Tectonic and Climate Controls. Journal of Coastal Research. 808-823.

Menendez, M., F.J. Mendez, I. Losada, and N.E. Graham, 2008. Variability of extreme wave heights in the northeast Pacific Ocean based on buoy measurements. Geophysical Research Letters, 35, p. L22607. http://dx.doi.org/10.1029/2008GL035394.

NOAA Ocean Service: http://oceanservice.noaa.gov/facts/population.html

NOAA Washington Coastal Management:
http://coastalmanagement.noaa.gov/mystate/wa.html

Rahmstorf, S., 2007. A semi-empirical approach to projecting future sea-level rise, Science, 315, 368-370.

Rahmstorf, S., 2010. A new view on sea level rise, Nature Geoscience, 4, 44-45.

Ruggiero, P., P.D. Komar, and J.C. Allan, 2010. Increasing wave heights and extreme-value projections: the wave climate of the U.S. Pacific Northwest, Coastal Engineering.

Ruggiero, P., C.A. Brown, P.D. Komar, J.C. Allan, D.A. Reusser, and H. Lee, 2010. Chapter 6: Impacts of climate change on Oregon's coasts and estuaries. Oregon Climate Assessment Report, K.D. Dello and P.W. Mote (eds). College of Oceanic and Atmospheric Sciences, Oregon State University, Corvallis, Oregon. Available at: www.occri.net/OCAR.

State of Oregon Coastal Management Program (OCMP):
http://www.oregon.gov/LCD/OCMP/CstZone_Intro.shtml

Washington State Department of Transportation (WSDOT):
http://www.wsdot.wa.gov/NR/rdonlyres/A2F8529C-32CC-4AE6-8CD7-FD459FCB6901/0/PNWRCEnvironmentalAssessmentAppendixGSeaLevelRise.pdf

Young, I.R., S. Zieger, and A.V. Babanin, 2011. Global trends in wind speed and wave height. Science, 332 (6028), 451-455. DOI:10.1126/science.1197219.

Likelihood of changes in abundance and geographical distributions of plant species and habitats for aquatic and terrestrial wildlife in the Northwest

Changes in the abundance and geographical distributions of plant species and habitats for aquatic and terrestrial wildlife in the Northwest.
Likelihood of occurrence by mid-21st century: High*

There is substantial evidence that climate change directly affects the abundance and geographical distributions of plant species (Chmura et al. 2011). The paleoenvironmental record demonstrates that plant species have responded individualistically to past climate changes (Davis and Shaw, 2001). Climate-driven changes in the abundance and geographical distribution of some plant species and habitat have been observed over the past century, for example, in changes in sub-alpine tree populations (e.g., lodgepole pine [*Pinus contorta*] in the central Sierra Nevada; Millar et al. 2004). Climate changes may affect plant phenology, such as plant flowering dates (e.g., common purple lilacs [*Syringa vulgaris* f. *purpurea*] and honeysuckle [*Lonicera* spp.] in the western United States, Cayan et al. 2001), which in turn may alter the timing and availability of plant resources used by other species. There is evidence that, for some species, plant responses to climate change may be mediated by the physiological response of plants to changes in atmospheric CO_2 concentrations and these responses may vary both within species and geographically across a species' range (Chmura et al. 2011).

In addition to being directly affected by changes in climate, plant species abundance and distribution in the Northwest also may be affected by climate-driven changes in disturbance regimes, such as wildfire (Littell et al. 2010), insect outbreaks (e.g., mountain pine beetle; Logan et al. 2003), disease (e.g., Swiss needle cast; Black et al. 2010), and drought occurrence (van Mantgem et al. 2009; Knutson and Pyke, 2008). The response of plant species to future climate changes may also be mediated by a number of other factors, including land use changes (e.g., grazing) and interactions with other species (e.g., invasive species) (Chambers and Wisdom, 2009). In the absence of empirical data for many species, various types of vegetation and species distribution models have been used to simulate Northwest plant species and habitat responses to potential future climate changes (e.g., Coops and Waring, 2011; Bachelet et al. 2011). In general, model simulations indicate large potential changes for some plant species and habitats in the Northwest, such as the simulated loss of subalpine habitat (Millar et al. 2006), although there are a number of uncertainties associated with model simulations of future vegetation change (McMahon et al. 2011).

Changes in the abundance and geographical distributions of aquatic and terrestrial wildlife.
Likelihood of occurrence by mid-21st century: High*

---

[9] Author: Sarah Shafer, U.S. Geological Survey, Corvallis, Oregon

As with plant species, there is substantial evidence that changes in climate may directly affect the abundance and geographical distributions of wildlife species (Root et al. 2003). Changes in the abundance and geographical distributions of aquatic and terrestrial wildlife have been documented for a number of species in the Northwest (e.g., Edith's checker-spot butterfly [*Euphydryas editha*], Parmesan, 2006; see Hixon et al. 2010; Janetos et al. 2008). Species may be directly affected by changes in climate (e.g., mortality from increased frequency of lethal temperatures) or indirectly affected by climate change effects on habitat (e.g., shifts in the seasonality and amounts of snowpack and runoff), disturbance regimes, competition and predator-prey interactions with other species, and disease (Parmesan, 2006; Hixon et al. 2010). Among species in the Northwest that have been identified as affected by the potential future loss of alpine and subalpine habitat are wolverines (*Gulo gulo*; Copeland et al. 2010) and pika (*Ochotona princeps*; Beever et al. 2010).

Wildlife species that are dependent on marine resources may be affected by climate change effects on marine systems, such as potential changes in the timing and strength of coastal ocean upwelling (e.g., Cassin's auklet [*Ptychoramphus aleuticus*]; Barth et al. 2007), gradual and abrupt changes in the distribution of sea surface temperatures (Payne et al. 2012), ocean acidification (Hofmann et al. 2010), the salinity of estuaries (see discussion in Ruggiero et al. 2010), and the occurrence of anoxic zones (Chan et al. 2008). A key Northwest issue identified in the 2009 "Global Climate Change Impacts in the US" report is the potential effect of future climate change on salmon and other aquatic species:

> **"Salmon and other coldwater species will experience additional stresses as a result of rising water temperatures and declining summer streamflows.** Northwest salmon populations are already at historically low levels due to variety of human-induced stresses. Climate change affects salmon throughout their life stages and poses an additional stress. Studies suggest that about a third of the current habitat for the Northwest's salmon and other coldwater fish will no longer be suitable for them by the end of this century due to climate change." (Northwest Fact Sheet; http://www.globalchange.gov /images/cir/region-pdf/NorthwestFactSheet.pdf)

Mantua et al. (2010) concluded that potential future climate changes could increase thermal stress for some salmon populations in Washington. Bull trout (*Salvelinus confluentus*) populations may also be affected by potential future climate changes (Rieman et al. 2007). As with plants, the response of aquatic and terrestrial wildlife to future climate changes may also be mediated by a number of other factors, including land use changes (e.g., grazing) and interactions with other species (e.g., invasive species) (Chambers and Wisdom, 2009).

Likelihood assessment based on paleoecological evidence of species' responses to past climate changes, observations of species' responses to climate changes over the past century, and empirical and modeling studies of species' sensitivities to changes in climate. Responses to future climate changes will vary by species depending on species sensitivities and potential future climate change.

Consequences:

Category	Consequences (**bold** denotes a positive consequence)
Natural Systems	• Species loss (ecotonal boundaries, local loss) – *(high)* • ESA Threatened species (e.g. sage grouse) • Species shifts (forest tree species highly vulnerable) • Ecological traps • Migration limited species (e.g. pika, pygmy rabbit, ponderosa pine: east of Cascades, but won't survive as they move west across the Cascades) – *(high)* • Ponderosa pine habitat wildlife (elk, white-headed woodpecker) – *(high)* • Salmon: hydrograph shift, earlier snowmelt, warm temperatures • Predator-prey dynamics (spatial redistribution) (e.g. as waters warm, mussels move deeper in sea where sea stars, which are predators of mussels, live) • Regional identity/keystone species • Increase in invasive species • Decoupling of ecosystems (e.g. pollinators & host plants) • Resources that provide habitat – *(high)* • Charismatic species – *(high)* • Changes in species abundance, productivity, phenology, timing, seasonality (see Case et al. 2010) • May have 2 algal bloom seasons in Puget Sound • **Increase in generalist species (e.g. coyotes, raccoons, crows)** • **Increase in exotic species** • **Increase in non-native species (e.g. American shad, small mouth bass)**
Managed Systems (e.g. Agriculture, Forestry)	• Shift of microclimates for wine varietals *(low)* • Crop yield changes
Built Environment	•
Economy	• Wine varietal microclimate sensitive to temperature – *(low)* • Loss of traditional animals & plants in place (tribes) • Crop yield changes • Resource based economies - *(high)* • **Wine industry in Oregon & Washington could see economic benefit of harboring microclimates able to support more wine grape varietals** • **Some species more abundant leading to economic opportunities or additional food source**
Human Society	• Seafood toxicity from shifting HAB species (may have 2 bloom seasons in Puget Sound) • Cultural effects of range shifts

| | • Tribal impacts (gathering traditions, first foods ceremonies, place-based limited treaty rights, identity)
• Changes in disease ecology
• Loss of livelihood if depend on certain species |

Human Population Affected: Very little direct effect on humans is expected from changing distributions of plant and animal species, as most of the comments indicated potential indirect effects. Respectively, 35.1%, 16.2%, and 27% of respondents indicated that few, about half, and most of the human population would be affected. It is difficult to quantify benefits that native species and habitats provide to human health and welfare. 21.6% of respondents indicated they didn't know. Some people feel that a large portion of the Northwest population is disconnected from nature and wouldn't even notice these kinds of changes. The definition of "affected" is unclear. Those at greatest risk are those who depend on and consume natural resources for a living, such as rural and native communities.

Human Mortality: It is unclear how human mortality would be affected by changes in distributions of plants and animals (91.8% of respondents indicated they didn't know, the sign was uncertain, or the question not applicable). However, changing distributions of disease vectors may relate to human disease incidence (8.1% of respondents indicated an increase in human mortality).

Human Health Quality: Although ecosystems provide goods and services essential to human health and welfare, it is uncertain how human health quality may be impacted (63.1% of respondents indicated they didn't know or the sign was uncertain). Possible consequences include enhanced allergies, psychological impacts for a region whose identity is partially tied to its ecosystems, food availability including public health implications for tribal members regarding first foods (especially salmon), and hardships from loss of natural resources used directly or for income. 28.9% of respondents indicated reduced general human health quality.

Biodiversity: There will be some winners and losers, but how species are affected will depend on the rate of change. Climate change may occur faster than some species can adapt spatially. Some shifts in distributions may be beneficial or neutral. Generally, the group feels that shifting distributions of plant and animal species due to climate change is unfavorable for biodiversity (44.7% of respondents indicated unfavorable, 28.9% indicated severe reduction, and 26.3% indicated they didn't know). Some unfavorable consequences may be that "weed" species generally outcompete native species or that there will be a loss of specialist species in favor of generalists. The pertinent theme from the respondent comments is the changing or disassociation of existing species assemblages as populations move to follow climate zones or are extirpated.

Geographic area affected: Most respondents (81.1%) indicated that most of the region would be affected by shifting distributions of plants and wildlife. The remainder of respondents indicated that a moderate portion of the region would be affected. Despite differences across microhabitats, changes to temperature and precipitation will likely affect all areas of the NW, and thus affect habitats and plant and animal species that respond to these climate conditions.

Built Environment: The potential disruption of public services and infrastructure due to changing distribution of plants and animals was indicated as minor (56.8% of respondents). The potential concern for disruption of infrastructure is the expansion of harmful invasive species that could potentially affect agriculture, grazing, aquaculture, and other infrastructure (13.5% of respondents indicated moderate or major infrastructure disruption).

Economy: Economies that could be impacted are tourism, especially wildlife-dependent tourism and recreation, natural resource economies like forest products, food production. Just fewer than half of respondents indicated they didn't know or the sign was uncertain.

Case Study Examples:
The wine grape industry in Oregon and Washington.

Special Considerations:
While the likelihood of shifting distributions of plant and animal species may be high since we are already seeing evidence of this occurring, the consequences are dependent on what the systems deliver and widely vary depending on the species under consideration. Shifting species distributions is dependent on many factors including climate, diseases, pathogens, and wildfire.

References:
Bachelet, D., B. R. Johnson, S. D. Bridgham, P. V. Dunn, H. E. Anderson and B. M. Rogers. 2011. Climate change impacts on western Pacific Northwest prairies and savannas. Northwest Science 85:411-429.
Barth, J. A., B. A. Menge, J. Lubchenco, F. Chan, J. M. Bane, A. R. Kirincich, M. A. McManus, K. J. Nielsen, S. D. Pierce, and L. Washburn. 2007. Delayed upwelling alters nearshore coastal ocean ecosystems in the northern California current. Proceedings of the National Academy of Sciences 104:3719-3724.
Beever, E. A., C. Ray, P. W. Mote, and J. L. Wilkening. 2010. Testing alternative models of climate-mediated extirpations. Ecological Applications 20:164-178.
Black, B. A., D. C. Shaw, and J. K. Stone. 2010. Impacts of Swiss needle cast on overstory Douglas-fir forests of the western Oregon Coast Range. Forest Ecology and Management 259:1673-1680.
Case, M. J., J. Lawler, J. Tomasevic, E. M. Gray, J. Langdon, B. McRae, J. Michael Scott, and L. Svancara. 2010. Assessing the sensitivities of species and ecosystems to climate change in the Pacific Northwest. PNW Climate Science Conference, Portland, Oregon, June 15-16, 2010.
Cayan, D. R., S. A. Kammerdiener, M. D. Dettinger, J. M. Caprio, and D. H. Peterson. 2001. Changes in the onset of spring in the western United States. Bulletin of the American Meteorological Society 82:399-415.

Chambers, J. C., and M. J. Wisdom. 2009. Priority research and management issues for the imperiled Great Basin of the western United States. Restoration Ecology 17:707-714.

Chan, F., J. A. Barth, J. Lubchenco, A. Kirincich, H. Weeks, W. T. Peterson, B. A. Menge. 2008. Emergence of anoxia in the California Current large marine ecosystem. Science 2008:920.

Chmura, D. J., P. D. Anderson, G. T. Howe, C. A. Harrington, J. E. Halofsky, D. L. Peterson, D. C. Shaw, and J. B. St. Clair. 2011. Forest responses to climate change in the northwestern United States: Ecophysiological foundations for adaptive management. Forest Ecology and Management 261:1121-1142.

Coops, N. C., and R. H. Waring. 2011. Estimating the vulnerability of fifteen tree species under changing climate in Northwest North America. Ecological Modeling, doi:10.1016/j.ecolmodel.2011.03.033

Copeland, J. P., K. S. McKelvey, K. B. Aubry, A. Landa, J. Persson, R. M. Inman, J. Krebs, E. Lofroth, H. Golden, J. R. Squires, A. Magoun, M. K. Schwartz, J. Wilmot, C. L. Copeland, R. E. Yates, I. Kojola, and R. May. 2010. The bioclimatic envelope of the wolverine (*Gulo gulo*): do climatic constraints limit its geographic distribution? Canadian Journal of Zoology 88:233-246.

Davis, M. B., and R. G. Shaw. 2001. Range shifts and adaptive responses to Quaternary climate change. Science 292:673-679.

Hixon, M. A., S. V. Gregory, W. D. Robinson. 2010. Oregon's fish and wildlife in a changing climate, in Oregon Climate Change Research Institute (K. D. Dello and P. W. Mote, eds.), Oregon Climate Assessment Report. College of Oceanic and Atmospheric Sciences, Oregon State University, Corvallis, OR. Available at: http://occri.net/ocar

Hofmann, G. E., J. P. Barry, P. J. Edmunds, R. D. Gates, D. A. Hutchins, T. Klinger, and M. A. Sewell. 2010. The effect of ocean acidification on calcifying organisms in marine ecosystems: an organism-to-ecosystem perspective. Annual Review of Ecology, Evolution, and Systematics 41:127-47.

Janetos, A., L. Hansen, D. Inouye, B. P. Kelly, L. Meyerson, B. Peterson, and R. Shaw. 2008. Biodiversity. In: The effects of climate change on agriculture, land resources, water resources, and biodiversity in the United States. A Report by the U.S. Climate Change Science Program and the Subcommittee on Global Change Research. Washington, DC, USA, 362 pp.

Knutson, K. C., and D. A. Pyke. 2008. Western juniper and ponderosa pine ecotonal climate-growth relationships across landscape gradients in southern Oregon. Canadian Journal of Forest Research, 38:3021-3032.

Littell, J. S., E. E. Oneil, D. McKenzie, J. A. Hicke, J. A. Lutz, R. A. Norheim, and M. M. Elsner. 2010. Forest ecosystems, disturbance, and climatic change in Washington State, USA. Climatic Change 102:129-158.

Logan, J. A., J. Régnière, and J. A. Powell. 2003. Assessing the impacts of global warming on forest pest dynamics. Frontiers in Ecology and the Environment 1:130-137.

Mantua, N., I. Tohver, and A. Hamlet. 2010. Climate change impacts on streamflow extremes and summertime stream temperature and their possible consequences for freshwater salmon habitat in Washington State. Climatic Change 102:187-223.

McMahon, S. M., S. P. Harrison, W. S. Armbruster, P. J. Bartlein, C. M. Beale, M. E. Edwards, J. Kattge, G. Midgley, X. Morin, and I. C. Prentice. 2011. Improving assessment and modeling of climate change impacts on global terrestrial biodiversity. Trends in Ecology and Evolution 26:249-259.

Millar, C. I., R. Neilson, D. Bachelet, R. Drapek, and J. Lenihan. 2006. Climate change at multiple scales. Chapter 3 in H. Salwasser and M. Cloughesy (eds). Forests, carbon, and climate change: a synthesis of science findings. Oregon Forest Resources Institute Publication.

Millar, C. I., R. D. Westfall, D. L. Delany, J. C. King, and L. J. Graumlich. 2004. Response of subalpine conifers in the Sierra Nevada, California, U.S.A., to 20th-century warming and decadal climate variability. Arctic, Antarctic, and Alpine Research 36:181-200.

Parmesan, C. 2006. Ecological and evolutionary responses to recent climate change. Annual Review of Ecology, Evolution, and Systematics 37:637-669.

Payne, M. C., C. A. Brown, D. A. Reusser, and H. Lee, II. 2012. Ecoregional analysis of nearshore sea-surface temperature in the North Pacific. PLoS ONE 7: e30105. doi:10.1371/journal.pone.0030105.

Rieman, B. E., D. Isaak, S. Adams, D. Horan, D. Nagel, C. Luce and D. Myers. 2007. Anticipated climate warming effects on bull trout habitats and populations across the interior Columbia River Basin. Transactions of the American Fisheries Society 136:1552-1565.

Root, T. L., J. T. Price, K. R. Hall, S. H. Schneider, C. Rosenzweig, and J. A. Pounds. 2003. Fingerprints of global warming on wild animals and plants. Nature 421:57-60.

Ruggiero, P., C. A. Brown, P. D. Komar, J. C. Allan, D. A. Reusser, and H. Lee, II. 2010. Impacts of climate change on Oregon's coasts and estuaries, in Oregon Climate Change Research Institute (K. D. Dello and P. W. Mote, eds.), Oregon Climate Assessment Report. College of Oceanic and Atmospheric Sciences, Oregon State University, Corvallis, OR. Available at: http://occri.net/ocar

van Mantgem, P. J., N. L. Stephenson, J. C. Byrne, L. D. Daniels, J. F. Franklin, P. Z. Fulé, M. E. Harmon, A. J. Larson, J. M. Smith, A. H. Taylor, and T. T. Veblen. 2009. Widespread increase of tree mortality rates in the western United States. Science 323:521-524.

Likelihood:[10]

Insects and diseases are key disturbances in the forest of the Northwest. Outbreaks of bark beetles and defoliating insects have affected millions of hectares of forest in the last several decades (USDA Forest Service, 2010; Hicke et al. 2012). Diseases also play an important role in regulating forest structure and function. Our general understanding that climate is a major driver of these biotic disturbances (Ayres and Lombardero, 2000; Bale et al. 2002; Raffa et al. 2008; Bentz et al. 2009; Sturrock et al. 2011). Climate influences forest insect and pathogen populations in several ways. Temperature directly affects insect mortality and life stage development rates. Unseasonably low temperatures during the fall, winter, and spring can kill insects (Wygant, 1940; Régnière and Bentz, 2007). Year-round temperatures regulate development rates, thereby influencing the number of years required to complete an insect's life cycle and, for bark beetles, affect population synchronization for mass-attacking host trees (Hansen et al. 2001; Logan and Powell, 2001). Pathogens are likewise impacted by temperature, and also influenced by foliar moisture (Sturrock et al. 2011). Biotic disturbances are indirectly governed by climate change effects on host tree stress, which is related to the capacity of a tree to defend itself from attack (Ayres and Lombardero, 2000; Sturrock et al. 2011). Drought events can weaken trees and have been linked to epidemics of biotic agents (Raffa et al. 2008), but major uncertainties remain.

Recent climate change has led to novel or more intense, frequent, or severe outbreaks of biotic disturbance agents in the Northwest and surrounding regions. Mountain pine beetle outbreaks have been influenced by warming in British Columbia, where northward expansion of the beetle is occurring (Safranyik et al. 2010), and in high-elevation forests in western North America, where warming has facilitated prolonged outbreaks in locations considered typically too cold to support the insect (Logan et al. 2010). Major outbreaks of two other bark beetles have been linked to warming and/or drought. Warm, dry summers were associated with a large epidemic of spruce beetle in Alaska (Berg et al. 2006; Sherriff et al. 2011). Extreme drought in the Southwest in the early 2000s was tightly coupled to a population increase of pinyon ips (Raffa et al. 2008). Expansion of area of Swiss needle cast, caused by *Phaeocryptopus gaeumannii*, a foliar pathogen that infests Douglas-fir in the coastal areas of the Northwest, has been associated with warming and precipitation changes (Stone et al. 2008). Sudden oak death, caused by *Phytophthora ramorum*, a virulent invasive pathogen in the Northwest, is affected by temperature and moisture (Venette and Cohen, 2006). Declines of yellow-cedar and aspen have been linked to warming and earlier snowmelt (Hennon et al. 2010) and drought stress (Worrall et al. 2010), respectively.

Based on our theoretical understanding of the role of climate, future climate change is expected to modify outbreaks of insects and diseases in the Northwest. In addition, specific projections have been made for some key biotic disturbance agents. The region of suitable temperatures for outbreaks of mountain pine beetle is predicted to move upslope with

---

[10] Author: Jeffrey A. Hicke, University of Idaho

future warming, continuing the high level of susceptibility of high-elevation pine forests to this insect (Littell et al. 2010). Similarly, the probability of one-year life cycles of spruce beetles will increase at high elevations in western North America, leading to enhanced probability of outbreak (Bentz et al. 2010). Climate change is predicted to increase the capacity of Swiss needle cast to infest Douglas-firs in the Pacific Northwest (Stone et al. 2008). Projections of climate change suggest increased impacts of sudden oak death in response to climate change (Sturrock et al. 2011).

Through alterations of forest structure and function, biotic disturbance agents have the capacity to affect future climate, thereby creating a feedback between climate change and insect and disease outbreaks (Adams et al. 2010; Hicke et al. 2012). Reductions in photosynthesis following attack and increases in decomposition of killed trees can result in forests becoming carbon sources instead of sink (Kurz et al. 2008). However, modifications to surface albedo can led to surface cooling that may be greater than warming associated with carbon release (O'Halloran et al. 2012).

Key outstanding gaps in knowledge include:
- improved understanding of the role of climate in most species, including the situations that result in more favorable or less favorable conditions
- development of monitoring system that ranges from plot to regional scales
- development of predictive models of insects and disease
- better understanding of interactions among disturbance agents and with other disturbances (windthrow, drought)

Consequences:

Category	Consequences

Natural Systems	• Pests/Agriculture Disease:
	  ○ Water quality decline from pesticides/polluted runoff
	• Forest Insects/Disease
	  ○ Bark beetle infestations
	  ○ Mortality of forest & vegetation species
	  ○ Increased wildfire risk
	  ○ Loss of forest services
	  ○ Loss of species dependent on old growth habitat (salmonids, migratory song birds, keystone wildlife: grizzly bear, wolverine, lynx)

Managed Systems (e.g. Agriculture, Forestry)	• Pests/Agriculture Disease:
	  ○ Potentially large impacts, wiped out: Wine, fruit trees, grass seed, wheat, nursery *(range from low to high)*
	  ○ East side – large scale farms: wheat industry
	  ○ West side – small scale farms & high value crops: horticulture, wine grapes, organic food, vegetables, grass seed, fruit trees
	  ○ Increased pesticide & fertilizer use/costs => increase carbon emissions

	o Livestock health risk
	• Forest Insects/Disease
	  o Forestry: bark beetle, pathogens, declines
	  o Reduced forest health (reduced growth rate, increased windthrow, increase vulnerability to invasive species)
	• Invasive species:
	  o Rangeland impacts from weeds in terms of land degradation – *(medium)*

Built Environment	•

Economy	• Pests/Agriculture Disease:
	  o Value of crops; cost of producing crops
	  o Farm income down, food prices up (potentially affecting regional/international markets)
	  o Small/Large farms impacts/vulnerabilities different
	  o Potential need to import more food
	• Human Disease – *(low to high, depending)*
	  o Shellfish industry decline (people won't eat oysters)
	• Forest Disease
	  o Forestry: pine bark beetle – *(high)*
	  o Decreased production/loss of commercially viable tree species
	  o Increase coast of pest & disease control

Human Society	• Pests/Agriculture Disease
	  o Increased use of pesticides & subsequent water quality decline => Food/water contamination
	  o Food increases impact the poor most
	• Human Disease – *(high, for limited pops)*
	  o Vector & water borne diseases (e.g. Dengue, West Nile Virus, Rift Valley Fever, Encephalitis, Hanta Virus, Cholera, San Joaquin River Valley Fever, Cryptococcus gattii from British Columbia) – *(high, for those with weak immune systems: children, elderly, diseased)*
	  o Communicable diseases – *(low, density dependent)*
	  o Food & water contamination
	  o Harmful Algal Blooms

Summary of Survey Rating Questions:

Human Population Affected: Humans could be indirectly affected by increases in insect outbreaks on forests insomuch as the fire potential increases. But, there are low population densities in affected areas, that is, mountainous areas. There are some diseases and pathogens affecting humans like West Nile and Lyme disease. 36.1% of respondents indicated that few people would be affected. One-fourth of respondents indicated either about half or most of the population would be affected. One-third indicated they didn't know.

Human Mortality: There is no clear direct effect on human mortality from diseases, invasive species, and pests (71.5% of respondents indicated they didn't know, the sign is uncertain, or was not applicable). Indirect effects include increased wildfire risks or water quality issues (28.6% indicated an increase in human mortality).

Human Health Quality: From forest insects and disease, the risk of wildfire may increase leading to human health quality consequences like reduced air quality and smoke inhalation. Increases in pathogens and toxic algae would also affect human health quality. 37.1% of respondents indicated reduced general human health quality. Pests to agricultural crops could increase pesticide use possibly resulting in water quality issues. 60% of respondents indicated they didn't know, the sign was uncertain or was not applicable.

Biodiversity: Generally, increases in diseases, insects, and pests were indicated as unfavorable for biodiversity (60% of respondents indicated unfavorable and 20% indicated severe reduction). Most comments were through the lens of forest insect outbreaks. Negative consequences from severe outbreaks could include shifts in food webs, impacts on "native" forests and wildlife, overtaken resiliency of systems, species may not be able to migrate to new environment if those areas killed by pests/pathogens. Positive consequences include possible increase in species richness because dead and dying forests provide resources for invertebrates, birds, and other species.

Geographic area affected: Over 80% of respondents indicated moderate or most of the region would be affected. In terms of forest disease and insect outbreaks, a large portion of the region is forestland and many areas are already seeing effects, mainly in Eastern Washington and Idaho. Most native conifers already have insects that are partially driven by climate, so likely to be affected as the climate changes.

Built Environment: Consequences of disease, insects, and pests were mostly indicated as either minor (47.2%) or moderate (11.1%), and relate mostly to fire impacts. There could be a loss of timber supply and impacts to water treatment facilities. Over one-fourth of respondents indicated they didn't know.

Economy: There is a potential for high economic costs due to loss of forest production and timber, and loss of agricultural revenue, food supply, and increased food costs due to crop failure (50% of respondents indicated high economic loss and 22.2% indicated medium loss). These economic losses are hard to quantify and there is low confidence to the extent that disease and pests will affect forests.

Identified Gaps in Knowledge:
Agricultural and vegetation impacts are not well captured in climate related models and hydrological impacts of forest mortality remain uncertain.

Special Considerations:

While the likelihood presentation focused on forest insect and disease outbreaks, the survey responses may include consequences of agricultural pests and diseases, human diseases, and other invasive species in addition to forest insects and diseases.

References:

Adams, H. D., A. K. Macalady, D. D. Breshears, C. D. Allen, N. L. Stephenson, S. R. Saleska, T. E. Huxman, and N. G. McDowell. 2010. Climate-induced tree mortality: Earth system consequences. EOS, Transactions of the American Geophysical Union **91**:153.

Ayres, M. P. and M. J. Lombardero. 2000. Assessing the consequences of global change for forest disturbance from herbivores and pathogens. The Science of the Total Environment **262**:263-286.

Bale, J. S., G. J. Masters, I. D. Hodkinson, C. Awmack, T. M. Bezemer, V. K. Brown, J. Butterfield, A. Buse, J. C. Coulson, J. Farrar, J. E. G. Good, R. Harrington, S. Hartley, T. H. Jones, R. L. Lindroth, M. C. Press, I. Symrnioudis, A. D. Watt, and J. B. Whittaker. 2002. Herbivory in global climate change research: direct effects of rising temperature on insect herbivores. Global Change Biology **8**:1-16.

Bentz, B., J. Logan, J. MacMahon, C. D. Allen, M. Ayres, E. Berg, A. Carroll, M. Hansen, J. Hicke, L. Joyce, W. Macfarlane, S. Munson, J. Negrón, T. Paine, J. Powell, K. Raffa, J. Régnière, M. Reid, B. Romme, S. J. Seybold, D. Six, D. Tomback, J. Vandygriff, T. Veblen, M. White, J. Witcosky, and D. Wood. 2009. Bark beetle outbreaks in western North America: Causes and consequences. University of Utah Press, Salt Lake City, UT.

Berg, E. E., J. D. Henry, C. L. Fastie, A. D. De Volder, and S. M. Matsuoka. 2006. Spruce beetle outbreaks on the Kenai Peninsula, Alaska, and Kluane National Park and Reserve, Yukon Territory: Relationship to summer temperatures and regional differences in disturbance regimes. Forest Ecology and Management **227**:219-232.

Hansen, M. E., B. J. Bentz, and D. L. Turner. 2001. Temperature-based model for predicting univoltine brood proportions in spruce beetle (Coleoptera: Scolytidae). Canadian Entomologist **133**:827-841.

Hennon, P. E., D. V. D'Amore, D. T. Witter, and M. B. Lamb. 2010. Influence of Forest Canopy and Snow on Microclimate in a Declining Yellow-cedar Forest of Southeast Alaska. Northwest Science **84**:73-87.

Hicke, J. A., C. D. Allen, A. R. Desai, M. C. Dietze, R. J. Hall, E. H. Ted Hogg, D. M. Kashian, D. Moore, K. F. Raffa, R. N. Sturrock, and J. Vogelmann. 2012. Effects of biotic disturbances on forest carbon cycling in the United States and Canada. Global Change Biology **18**:7-34.

Kurz, W. A., C. C. Dymond, G. Stinson, G. J. Rampley, E. T. Neilson, A. L. Carroll, T. Ebata, and L. Safranyik. 2008. Mountain pine beetle and forest carbon feedback to climate change. Nature **452**:987-990.

Logan, J. A., W. W. Macfarlane, and L. Willcox. 2010. Whitebark pine vulnerability to climate-driven mountain pine beetle disturbance in the Greater Yellowstone Ecosystem. Ecological Applications **20**:895-902.

Logan, J. A. and J. A. Powell. 2001. Ghost forests, global warming and the mountain pine beetle (Coleoptera: Scolytidae). American Entomologist **47**:160-173.

O'Halloran, T. L., B. E. Law, M. L. Goulden, Z. Wang, J. G. Barr, C. Schaaf, M. Brown, J. D. Fuentes, M. Göckede, A. Black, and V. Engel. 2012. Radiative forcing of natural forest disturbances. Global Change Biology **18**:555-565.

Raffa, K. F., B. H. Aukema, B. J. Bentz, A. L. Carroll, J. A. Hicke, M. G. Turner, and W. H. Romme. 2008. Cross-scale drivers of natural disturbances prone to anthropogenic amplification: The dynamics of bark beetle eruptions. BioScience **58**:501-517.

Régnière, J. and B. Bentz. 2007. Modeling cold tolerance in the mountain pine beetle, *Dendroctonus ponderosae*. Journal of Insect Physiology **53**:559-572.

Safranyik, L., A. L. Carroll, J. Regniere, D. W. Langor, W. G. Riel, T. L. Shore, B. Peter, B. J. Cooke, V. G. Nealis, and S. W. Taylor. 2010. Potential for range expansion of mountain pine beetle into the boreal forest of North America. Canadian Entomologist **142**:415-442.

Sherriff, R. L., E. E. Berg, and A. E. Miller. 2011. Climate variability and spruce beetle (Dendroctonus rufipennis) outbreaks in south-central and southwest Alaska. Ecology **92**:1459-1470.

Stone, J. K., L. B. Coop, and D. K. Manter. 2008. Predicting effects of climate change on Swiss needle cast disease severity in Pacific Northwest forests. Canadian Journal of Plant Pathology **30**:169-176.

Sturrock, R. N., S. J. Frankel, A. V. Brown, P. E. Hennon, J. T. Kliejunas, K. J. Lewis, J. J. Worrall, and A. J. Woods. 2011. Climate change and forest diseases. Plant Pathology **60**:133-149.

USDA Forest Service. 2010. Major Forest Insect and Disease Conditions in the United States: 2009 Update. FS-952, Washington, D.C.

Venette, R. C. and S. D. Cohen. 2006. Potential climatic suitability for establishment of *Phytophthora ramorum* within the contiguous United States. Forest Ecology and Management **231**:18-26.

Worrall, J. J., S. B. Marchetti, L. Egeland, R. A. Mask, T. Eager, and B. Howell. 2010. Effects and etiology of sudden aspen decline in southwestern Colorado, USA. Forest Ecology and Management **260**:638-648.

Wygant, N. D. 1940. Effects of low temperature on the Black Hills beetle (*Dendroctonus ponderosae* Hopkins). Ph.D. dissertation. State College of New York, Syracuse, NY.

Likelihood:[11]

Observed changes in extreme precipitation during the past several decades are ambiguous. The general picture that emerges from several studies using different definitions of extremes is that in Washington State, especially western Washington, most metrics of extremes show increases of 10-20%. In the rest of the region, results are more mixed, with some locations, time periods, and metrics showing increases and others decreases.

Projected future changes in extreme precipitation are less ambiguous (Table 2). The NARCCAP results indicate increases throughout the Northwest in the number of days > 1".

Table 2. Mean changes, along with the standard deviation of selected precipitation variables from the NARCCAP simulations.

Metric of extreme precipitation	NARCCAP Mean Change	NARCCAP St. Dev. Of Change
#days> 1 inch	+13%	7%
#days > 2 inches	+15%	14%
#days > 3 inches	+22%	22%
#days > 4 inches	+29%	40%
Max run days <0.1 inches	+6 days	+3 days

Consequences:

Category	Consequences
Natural Systems	• Decreased water quality from storm water and combined storm water-sewer overflows puts untreated waste water into streams and other waterways • Increased turbidity from erosion & landslides (e.g. Willamette River) affects fish in some rivers – *(high)* • Loss of habitat for spawning salmon and salmonids from scouring floods • Changes in stream hydrology & physical characteristics • Tree & forest mortality from windthrow events due to unstable slopes and saturated soils • Impacts on coastal estuarine circulation and wetlands from freshwater input • Increased TDML & delivery of sediments, nutrients, & contaminants to Puget Sound • Increased risk of landslides • Debris flow cause flooding • Losing glacier buffering (e.g. Rainier)

---

[11] Author: Philip Mote, Oregon State University

Managed Systems (e.g. Agriculture, Forestry)	• Dam management trade off between water supply and flood risk • Crop loss/reduce production • Soil loss/erosion of farmlands • Livestock loss • Reservoir systems in danger • Decline shellfish production
Built Environment	• Flood & landslide damage (roads, including forest access roads, culverts, bridges, power substations, water supply infrastructure) – *(high)* • Erosion & extreme precipitation events detrimental to dam operations & energy infrastructure • Increased loss & damages to homes & property in low-lying areas & floodplains • Disrupt transportation networks • Urban flooding • Increased turbidity affect municipalities with no backup well field – *(high)*
Economy	• Flood and debris flow damages and repair costs to infrastructure such as road, bridges, and culverts • Reduction in agricultural income
Human Society	• Landslide & flood hazards increase potential deaths and injuries • Decreased water quality from storm water overflows affecting drinking water and human health • Water borne illnesses • Respiratory problems from increased household mold • Impaired recreation from debris flow (e.g. Salmon River)

Summary of Survey Rating Questions:

Human Population Affected: The same number of respondents (32.4% each) indicated that either about half or most of the population would be affected. About one-fifth indicated that few people would be affected. The human population will likely experience consequences directly and/or indirectly. Direct consequences from floods would impact most population centers in the Northwest, especially those built on rivers and in floodplains, urban populations and isolated populations. Indirect effects include contaminated water supply, economic impacts, and transportation disruption (e.g. I-5 closure). Consequences will vary by the extent of the event.

Human Mortality: There is a potential for increases in human mortality (as indicated by 48.6% of respondents) due to extreme precipitation and flooding. Possible deaths associated with the risks of increased flooding include disease from standing water and septic system failures, although good water quality monitoring may likely limit the health impacts of contaminated water supply. The number of flood deaths is quite small, and are often not fatal owing to good flood predictions. Increases in automobile accidents from extreme precipitation events and associated deaths may occur. Much uncertainty exists as roughly half of respondents indicated they didn't know or the sign was uncertain.

Human Health Quality: There is potential for reduced human health quality (indicated by 38.9% of respondents) due to water borne diseases, contamination of water supply and water quality, potential for injury, and if drinking and waste water facilities are out of service. Over half of respondents indicated they didn't know or the sign was uncertain.

Biodiversity: Extreme precipitation and flooding events may be unfavorable to biodiversity (indicated by 41.7% of respondents) due to damage, disruption, and loss of habitat from erosion, flooding and debris flows. There may be a short-term decrease in biodiversity, but in the long run, there is a potential increase in biodiversity on riverine systems unless the floods occur too frequently. Half the respondents indicated they didn't know. Flooding also carries a positive benefit to ecosystems, transporting sediments, nutrients, and habitat-promoting debris.

Geographic area affected: Respondents indicated that more frequent extreme precipitation and flooding events would affect a moderate portion (54.1% of respondents) to most (29.7% of respondents) of the geographic area of the Northwest, mostly west of the Cascades. Areas particularly vulnerable to flooding include those near rivers, streams, and wetland and adjacent floodplains. The frequency and severity of future extreme precipitation is uncertain and a major factor involves changes in storm tracks and extreme events associated with atmospheric rivers.

Built Environment: Respondents indicated that increases in extreme precipitation and flooding would have major (48.6%), moderate (29.7%), or minor (16.2%) consequences for public services and infrastructure, including roads and rails (e.g. I-5 closure as precedent), storm water and waster water treatment plants, urban areas, and levees and dams. The consequences to the built environment could be considerable given that communities tend to build infrastructure near rivers and in floodplains. The consequences depend a lot on the magnitude of change of extreme precipitation events.

Economy: Respectively, 45.9% and 24.3% of respondents indicated high and medium economic loss. According to comments, the main systems affected are transportation networks, agricultural systems (from soil loss), and other buildings and utilities.

Case Study Examples:
The rain-on-snow flood events of 1996 in Oregon resulted in a disaster declaration for three-quarters of the counties. (Halpert and Bell, 1996; Chang et al. 2010; State of Oregon report).

It was also mentioned that adaptation efforts are already underway in terms of infrastructure investments to elevate above "flashy" creeks.

Special Considerations:

It was noted that there are contrasting concerns, with more extreme precipitation concentrated in winter and decreasing seasonal precipitation in summer. Flooding can result from extreme precipitation, high temperatures melting snowpack, or a combination of both known as rain on snow events. It was also noted that runoff doesn't scale linearly with extreme precipitation. It is important to distinguish between small urban watersheds and larger rivers. In small watersheds, floods are controlled by precipitation intensities over a timescale of hours. Larger river systems respond to longer time periods including antecedent conditions such as snowpack leading to rain on snow events. In snow-dominated basins, the snowline during extreme precipitation events partially determines the runoff response.

References:

Halpert, M. S., and G. D. Bell, 1997: Climate assessment for 1996. *Bull. Amer. Meteor. Soc.,* **78,** S1–S49. http://dx.doi.org/10.1175/1520-0477(1997)078%3C1038:CAF%3E2.0.CO;2

Heejun Chang, Martin Lafrenz, Il-Won Jung, Miguel Figliozzi, Deena Platman & Cindy Pederson (2010): Potential Impacts of Climate Change on Flood-Induced Travel Disruptions: A Case Study of Portland, Oregon, USA, Annals of the Association of American Geographers, 100:4, 938-952. http://dx.doi.org/10.1080/00045608.2010.497110

State of Oregon Legislative, Policy, Research, and Committee Services report: Basics on Floods in Oregon: http://www.leg.state.or.us/comm/commsrvs/floods

CONCLUSIONS

The aspirational goal for this workshop was to develop an initial ranking of each risk in terms of likelihood and magnitude of consequences. The final group plenary discussion was designed to discuss and defend the placement of all the risks together on a common risk matrix resulting in a preliminary prioritization of key risks and vulnerabilities with associated rationale from the groups. While the actual course of the workshop deviated from the initial plan, the breakout group discussions were engaging and productive and many points for consideration in implementing this risk-based framing were discussed. We also describe below the lessons learned throughout the workshop from the design phase through implementation.

Lessons Learned

This workshop was characterized by a fast-paced, intensive morning and a productive, engaging afternoon. The workshop participants spent the morning in a sequence of events alternating between listening to a five-minute presentation on a single climate risk and spending five to ten minutes furiously typing in their survey answers. The afternoon was spent in deep discussion. Some commented that a five-minute presentation was too short to convey the background, and none of the groups had time to discuss more than one or two of the breakout group charges. A two-day, rather than single-day, workshop might have been better suited to accommodate the goals and components of this workshop.

The use of the online survey was an efficient and effective tool to collect a wealth of opinions, comments, and rationale from local experts. There are several ways in which the design, implementation, and use of survey results could be improved and are summarized as a list of recommendations. First, a team of physical and social science and communications experts should be consulted when designing the survey. Some of the questions were difficult to answer because responses depended on various factors. There was some concern about the appropriate knowledge and time for providing answers to some of the rating questions. Second, to make better use of the survey results during the workshop, the survey could have been distributed and results compiled before the workshop. One challenge of the breakout groups was getting beyond compiling a list of consequences and ranking those consequences. The survey asked participants to list all the consequences of the risk. Designing more structure into the survey so as to categorize the consequences may have better facilitated the use of the open-ended survey results and advanced the breakout group conversations toward ranking and rationales instead of spending as much time recapitulating the consequences. In general, the surveys were successful in that everyone was able to provide their opinions and those opinions were summarized into the previous risk summaries.

Recommendations:
- Consider setting aside a day and a half or two days for a similar workshop
- Iterate with a team of physical and social science and communications expert when designing a similar survey
- Consider distributing a survey and compiling survey results before the workshop

- Structure a survey in such a way that the results feed directly into the next phase of the group discussions

A number of challenges arose during the workshop. It was difficult for groups to talk about consequences without bringing in likelihood. In many cases, the likelihood of a particular risk occurring is quite uncertain. Some groups felt that the consequences depended on the likelihood of a risk occurring, thus much of the breakout group discussion tended toward characterizing the likelihood. It was not defined ahead of time how to think about the dependence of consequences on the likelihood, but an impromptu suggestion was to consider the consequences assuming the risk occurs. The groups highlighted the fact that the Northwest region is characterized by heterogeneous geographic texture, and many participants were reluctant to make sweeping generalizations about the entire region. Another challenge in some groups was the hesitancy to qualitatively rank consequences.

Challenges:
- Separation of likelihood and consequences
- Considering the Northwest region generally while recognizing geographic texture
- Defining a system for qualitative ranking

Despite these challenges, group discussions were quite productive. Some groups were able to begin to characterize by risk ranking and other groups suggested other ways of talking about risks. All groups brought up many common issues and points of consideration. One common issue was how to incorporate adaptive capacity in the determination of magnitude of consequences. Without adaptive measures, the consequences of some risks could be quite large, but could easily be remedied with relatively easy and common adaptation strategies such as use of air conditioning during extreme heat events to prevent human mortality. Another common issue was determining an aggregate magnitude of consequences, when different sub-regions and watersheds are dominated by different climate impacts. One suggestion was to disaggregate the risk assessment either by the eastern and western parts of the region, or consider separate risks to watersheds with and without abundant storage. Many participants commented about the need to incorporate a way to include the positive opportunities in addition to the negative consequences on the risk matrix. In terms of developing the Northwest chapter of the NCA following this workshop, several participants suggested incorporating short compelling narratives to get a flavor of risks and vulnerabilities in the Northwest.

Considerations:
- Magnitude of consequences depends on adaptive capacity
- Sub-regional disaggregation of risks
- Include positive consequences within the risk matrix
- Use of short compelling narratives to tell the story of climate risks and vulnerabilities in the Northwest

REFERENCES

Climate Impacts Group, 2009. The Washington Climate Change Impacts Assessment, M. McGuire Elsner, J. Littell, and L Whitely Binder (eds). Center for Science in the Earth System, Joint Institute for the Study of the Atmosphere and Oceans, University of Washington, Seattle, Washington. Available at: http://www.cses.washington.edu/db/pdf/wacciareport681.pdf

Intergovernmental Panel on Climate Change (IPCC), *Climate Change 2007a: Synthesis Report,* Cambridge University Press, Cambridge, 2007.

Intergovernmental Panel on Climate Change (IPCC), Climate Change 2007b: Synthesis Report, Summary for Policy Makers. Cambridge University Press, Cambridge, 2007.

Morsch, Amy, and Kathryn Saterson, 2010. A Climate Change Vulnerability and Risk Assessment for the City of Atlanta, Georgia. Master's project submitted to the Nicholas School of the Environment of Duke University. April 2010.

National Climate Assessment strategic plan summary: http://www.globalchange.gov/what-we-do/assessment/backgroundprocess/strategic-plan

National Research Council (NRC), *Adapting to the Impacts of Climate Change,* Report of the Panel on Adapting to the Impacts of Climate Change, America's Climate Choices, 2010a, http://www.nas.edu.

National Research Council (NRC), *Informing Decisions on Climate Change,* Report of the Panel on Informing Decisions on Climate Change, America's Climate Choices, 2010b, http://www.nas.edu.

National Research Council (NRC), *Report of the Committee on Americas Climate Choices*, 2011, http://www.nas.edu.

New York City Panel on Climate Change (NPCC), 2010, *Climate Change Adaptation in New York City: Building a Risk Management Response.* C. Rosenzweig & W. Solecki, Eds, prepared for use by the New York City Climate Change Adaptation Task Force, *Annals of the New York Academy of Sciences,* New York, NY. 349 pp., http://www.nyas.org.

Oregon Climate Change Adaptation Framework (2010). State of Oregon. http://www.oregon.gov/LCD/docs/ClimateChange/Framework_Final.pdf?ga=t

Oregon Climate Change Research Institute (2010), Oregon Climate Assessment Report, K.D. Dello and P.W. Mote (eds). College of Oceanic and Atmospheric Sciences, Oregon State University, Corvallis, Oregon. Available at: www.occri.net/OCAR.

Snover, A.K., L. Whitely Binder, J. Lopez, E. Willmott, J. Kay, D. Howell, and J. Simmonds, 2007. Preparing for Climate Change: A Guidebook for Local, Regional, and State Governments. In association with and published by ICLEI-Local Governments for Sustainability, Oakland, CA.

US Global Change Research Program (USGCRP): http://www.globalchange.gov/

Yohe, G. and Tol, R., "Indicators for Social and Economic Coping Capacity – Moving Toward a Working Definition of Adaptive Capacity", *Global Environmental Change* **12**: 25-40, 2002.

Yohe, G., "Risk Assessment and Risk Management for Infrastructure Planning and Investment", *The Bridge* **40**(3): 14-21, National Academy of Engineering, 2010.

APPENDIX A: WORKSHOP PARTICIPANTS & CONTRIBUTORS

Workshop Attendees & Participants:

+Presenter; #Facilitator; *Coauthor; %Reviewer; &Organizer; >Remote participant

Hedia Adelsman
Washington Department of Ecology
P.O. Box 47600
Olympia, WA 98504-7600
Phone: (360) 407-6222
Fax: (360) 407-6989
E-mail: hade461@ecy.wa.gov

John Antle
Agricultural & Resource Economics
Oregon State University
213 Ballard Extension Hall
Corvallis, OR 97331-3601
Phone: (541) 737-1425
Fax: (541) 737-2563
E-mail: john.antle@oregonstate.edu

Virginia Armbrust (&)
School of Oceanography
University of Washington
Box 357940
Seattle, WA 98195-7940
Phone: (206) 616-1783
E-mail: armbrust@u.washington.edu

Jeffrey Bethel
College of Public Health & Human Sciences
Oregon State University
401 Waldo Hall
Corvallis, OR 97331
Phone: (541) 737-3832
Fax: (541) 737-4001
E-mail: jeff.bethel@oregonstate.edu

Cheryl Brown
US Environmental Protection Agency
Pacific Coastal Ecology Branch
2111 S.E. Marine Science Center Drive
Newport, OR 97365
Phone: (541) 867-4042
Fax: (541) 867-4049
E-mail: brown.cheryl@epa.gov

Susan Capalbo
Agricultural & Resource Economics
Oregon State University
213 Ballard Extension Hall
Corvallis, OR 97331-3601
Phone: (541) 737-5639
Fax: (541) 737-1441
E-mail: susan.capalbo@oregonstate.edu

Meghan Dalton (&#*%)
Oregon Climate Change Research Institute
College of Earth Ocean and Atmospheric Sciences
Oregon State University
326 Strand Hall
Corvallis, OR 97331
Phone: (541) 737-3081
Fax: (541) 737-2540
E-mail: mdalton@coas.oregonstate.edu

Kathie Dello (&#+%)
Oregon Climate Change Research Institute
College of Earth Ocean and Atmospheric Sciences
Oregon State University
326 Strand Hall
Corvallis, OR 97331
Phone: (541) 737-8927
Fax: (541) 737-2540
E-mail: kdello@coas.oregonstate.edu

Bill Drumheller
Oregon Department of Energy
625 Marion St. NE
Salem, OR 97301-3737
Phone: (503) 378-4035
E-mail: bill.drumheller@state.or.us

Julie Early-Alberts
Oregon Public Health Division
Oregon Health Authority
800 NE Oregon St., Suite 640
Portland, OR 97232-2162
Phone: (971) 673-0438
Fax: (971) 673-0979
E-mail: Julie.early-alberts@state.or.us

Jason Eisdorfer
Bonneville Power Administration
P.O. Box 3621
Portland, OR 97208-3621
Phone: (503) 230-3589
E-mail: jgeisdorfer@bpa.gov

Sean Finn
Great Northern Landscape Conservation
Cooperative
US Fish & Wildlife Service
Snake River Field Station
970 Lusk St.
Boise, ID 83706
Phone: (208) 426-2697
E-mail: sean_finn@fws.gov

Josh Foster (&#%)
Oregon Climate Change Research Institute
College of Earth Ocean and Atmospheric Sciences
Oregon State University
326 Strand Hall
Corvallis, OR 97331
Phone: (541) 737-5262
Fax: (541) 737-2540
E-mail: jfoster@coas.oregonstate.edu

Laura Gephart
Columbia River Inter-Tribal Fish Commission
729 N.E. Oregon St, Suite 200
Portland, OR 97232
Phone: (503) 238-0667
Fax: (503) 235-4228
E-mail: gepl@critfc.org

Patty Glick
National Wildlife Federation
6 Nickerson Street, Suite 200
Seattle, WA 98109
Phone: (206) 577-7825
E-mail: glick@nwf.org

Gordon Grant
USDA Forest Service
Geosciences
Oregon State University
104 Wilkinson Hall
Corvallis, OR 97331
Phone: (541) 750-7328
Fax: (541) 750-7329
E-mail: Gordon.grant@oregonstate.edu

Andrea Hamberg
Oregon Public Health Division
Oregon Health Authority
800 N.E. Oregon St., Suite 640
Portland, OR 97232-2162
Phone: (971) 673-0444
Fax: (971) 673-0979
E-mail: andrea.hamberg@state.or.us

Preston Hardison
Tulalip Tribes of Washington
Natural Resource Department
7515 Totem Beach Road
Tulalip, WA 98271
Phone: (360) 716-4480
Fax: (360) 651-4490
E-mail: prestonh@comcast.net

Jeffrey Hicke (>+*%)
Geography
University of Idaho
P.O. Box 443021
Moscow, ID 83844-3021
Phone: 208-885-6240
Fax: 208-885-2855
E-mail: jhicke@uidaho.edu

Meg Jones (+&)
Washington State Office of the Insurance
Commissioner
Insurance Building, Capitol Campus
Olympia, WA 98504
Phone: (360) 725-7170
E-mail: megj@oic.wa.gov

Beverly Law
Forest Ecosystems & Society
Oregon State University
321 Richardson Hall
Corvallis, OR 97331-5752
Phone: (541) 737-6111
Fax: (541) 737-1393
E-mail: bev.law@oregonstate.edu

Dennis Lettenmaier (+*)
CSES Climate Impacts Group
University of Washington
Box 355672
Seattle, WA 98195-5672
Phone: (206) 543-2532
E-mail: dennisl@uw.edu

Heather Lintz
Oregon Climate Change Research Institute
College of Earth Ocean and Atmospheric Sciences
Oregon State University
326 Strand Hall
Corvallis, OR 97331
Phone: (541) 737-2996
Fax: (541) 737-2540
E-mail: hlintz@coas.oregonstate.edu

Fred Lipschultz (&)
US Global Change Research Program
1717 Pennsylvania Ave NW, Suite 250
Washington, D.C. 20006
Phone: (202) 419-3463
E-mail: flipschultz@usgcrp.gov

Jeremy Littell (+*)
CSES Climate Impacts Group
University of Washington
Box 355672
Seattle, WA 98195-5672
Phone: (206) 221-2997
E-mail: jlittell@uw.edu

Kathy Lynn
Environmental Studies Program
University of Oregon
243 Columbia Hall
Eugene, OR 97403-5295
Phone: (541) 346-5777
E-mail: Kathy@uoregon.edu

Mary Mahaffy
North Pacific Landscape Conservation Cooperative
US Fish & Wildlife Service
510 Desmond Drive SE, Suite 102
Lacey, WA 98503-1263
Phone: (360) 753-7763
E-mail: mary_mahaffy@fws.gov

Nathan Mantua
Climate Impacts Group
JISAO Center for Science in the Earth System
University of Washington
Box 355672
Seattle, WA 98195-4235
Phone: (206) 616-5347
E-mail: nmantua@u.washington.edu

Stephanie Moore
NOAA Northwest Fisheries Science Center
2725 Montlake Blvd E
Seattle, WA 98112
Phone: (206) 860-3327
Fax: (206) 860-3335
E-mail: Stephanie.moore@noaa.gov

Hamid Moradkhani
Department of Civil & Environmental Engineering
Portland State University
1930 S.W. 4th Avenue, Suite 200
Portland, OR 97201
Phone: (503) 725-2436
Fax: (503) 725-5950
E-mail: hamidm@cecs.pdx.edu

Philip Mote (+#*&%)
Oregon Climate Change Research Institute
College of Earth Ocean and Atmospheric Sciences
Oregon State University
326 Strand Hall
Corvallis, OR 97331
Phone: (541) 737-5694
Fax: (541) 737-2540
E-mail: pmote@coas.oregonstate.edu

Phillip Pasteris
Global Water Resources
CH2M Hill
2020 SW 4th Avenue
Portland, OR 97201
Phone: (503) 736-4301
E-mail: phillip.pasteris@ch2m.com

David Patte
US Fish & Wildlife Service
Pacific Region
911 N.E. 11th Avenue
Portland, OR 97232
Phone: (503) 231-6120
E-mail: david_patte@fws.gov

Rick Raymondi (%)
Idaho Department of Water Resources
322 East Front Street
P.O. Box 83720
Boise, ID 83720-0098
Phone: (208) 287-4800
Fax: (208) 287-6700
E-mail: rick.raymondi@idwr.idaho.gov

Spencer Reeder
Cascadia Consulting Group
1109 First Avenue, Suite 400
Seattle, WA 98101
Phone: (206) 343-9759 ext. 102
Fax: (206) 343-9819
E-mail: spencer@cascadiaconsulting.com

Peter Ruggiero (>+*)
Geosciences
Oregon State University
104 Wilkinson Hall
Corvallis, OR 97331-5506
Phone: (541) 737-1239
Fax: (541) 737-1200
E-mail: ruggierp@science.oregonstate.edu

John Schweiss
US Environmental Protection Agency Region 10
1200 Sixth Avenue
Seattle, WA 98101
Phone: (206) 553-1690
Fax: (206) 553-1809
E-mail: schweiss.jon@epa.gov

Michael Scott
Energy and Environment Directorate
Pacific Northwest National Laboratory
902 Battelle Boulevard
P.O. Box 999, MSIN K6-05
Richland, WA 99352
Phone: (509) 372-4273
Fax: (509) 372-4370
E-mail: Michael.scott@pnnl.gov

Sarah Shafer (+*%)
Geology & Environmental Change Science Center
US Geological Survey
3200 SW Jefferson Way
Corvallis, OR 97331
Phone: (541) 750-0946
Fax: (541) 750-0969
E-mail: sshafer@usgs.gov

Amy Snover
Climate Impacts Group
JISAO Center for Science in the Earth System
University of Washington
Box 355672
Seattle, WA 98195
Phone: (206) 221-0222
Fax: (206) 616-5775
E-mail: aksnover@uw.edu

John Stevenson (#&)
Oregon Climate Change Research Institute
College of Earth Ocean and Atmospheric Sciences
Oregon State University
326 Strand Hall
Corvallis, OR 97331
Phone: (541) 737-5689
Fax: (541) 737-2540
E-mail: jstevenson@coas.oregonstate.edu

Lorna Stickel
Portland Water Bureau
1120 SW Fifth Avenue, 6th Floor
Portland, OR 97204
Phone: (503) 823-7404
E-mail: lorna.stickel@portlandoregon.gov

Claudio Stöckle
Biological Systems Engineering
Washington State University
P.O. Box 64120
Pullman, WA 99164-6120
Phone: (509) 335-1578
Fax: (509) 335-2722
E-mail: stockle@wsu.edu

Toni Turner
Pacific Northwest Regional Office
US Bureau of Reclamation
1150 North Curtis Road, Suite 100
Boise, ID 83706-1234
Phone: (208) 378-5025
E-mail: tturner@usbr.gov

Beatrice Van Horne
Pacific Northwest Research Station
US Forest Service
Corvallis, OR
Phone: (541) 750-7357
E-mail: bvhorne@fs.fed.us

Stacy Vynne
The Resource Innovation Group
Climate Leadership Initiative
P.O. Box 51182
Eugene, OR 97405
Phone: (541) 654-4048
E-mail: stacy@trig-cli.org

Jeff Weber
Oregon Coastal Management Program
Oregon Department of Land Conservation &
Development
800 N.E. Oregon St. #18, Suite 1145
Portland, OR 97232
Phone: (971) 673-0964
Fax: (971) 673-0911
E-mail: jeff.weber@state.or.us

Gary Yohe (>+)
Economics & Environmental Studies
Wesleyan University
238 Church Street
Middletown, CT 06459
Phone: (860) 685-3658
E-mail: gyohe@wesleyan.edu

Andrew Yost
Oregon Department of Forestry
2600 State Street
Salem, OR 97310
Phone: (503) 945-7410
E-mail: ayost@odf.state.or.us

Workshop Contributors:

Paul Fleming (&)
Seattle Public Utilities
700 5th Avenue, Suite 4900
P.O. Box 34018
Seattle, WA 98124-4018
Phone: (206) 684-7626
E-mail: paul.fleming@seattle.gov

Denise Lach (&)
Sociology Department
Oregon State University
307 Fairbanks Hall
Corvallis, OR 97331-3703
Phone: (541) 737-5471
Fax: (541) 737-5372
E-mail: denise.lach@oregonstate.edu

Ron Mitchell (&)
Political Science
University of Oregon
921 PLC
Eugene, OR 97403-1284
Phone: (541) 346-4880
E-mail: rmitchel@uoregon.edu

Jan Newton (*%)
Applied Physics Laboratory
University of Washington
1013 N.E. 40th Street
Seattle, WA 98105-6698
Phone: (206) 543-9152
Fax: (206) 543-6785
E-mail: newton@ocean.washington.edu

Terese Neu Richmond (&)
GordonDerr LLP
2025 First Avenue, Suite 500
Seattle, WA 98121-3140
Phone: (206) 382-9540
E-mail: trichmond@gordonderr.com

Organizations Represented:

North Pacific Land Conservation Cooperative
Portland Water Bureau
Columbia River Inter-Tribal Fish Commission
Washington State Department of Ecology
Great Northern Land Conservation Cooperative
Idaho Department of Water Resources
Oregon Department of Land Conservation and Development
Portland State University
Oregon Climate Change Research Institute
National Wildlife Federation
US Fish and Wildlife Service
US EPA Region 10
Tulalip Tribes of Washington
Pacific Northwest National Laboratory
Bonneville Power Administration
US Global Change Research Program
Oregon State University
Oregon Department of Energy
NOAA Northwest Fisheries Science Center
US Geological Survey
Washington State University
Cascadia Consulting Group
University of Oregon
Pacific Northwest Tribal Climate Change Network
The Resource Innovation Group
University of Washington
USDA Forest Service Pacific Northwest Research Station
Oregon Department of Forestry
CH2M Hill
US EPA Office of Research & Development
US Bureau of Reclamation
Oregon Health Authority

APPENDIX B: SURVEY QUESTIONS

Introduction:
This survey will ask you for your expert judgment n the likelihood and consequences of climate risks in the Northwest region, defined as Oregon, Washington, & Idaho. The survey is designed so that you keep pace with the series of presentations introducing climate risks and the likelihood of occurrence. Following each presentation, you will be asked to answer a few questions pertaining to the risk that was just presented. Please follow along with the pace of the presentations. Selected results from this survey will be available as a tool for the breakout session groups. There are 9 pages in total: a page of pre-survey questions, and a page for each risk. Be sure to click "Next" directly after completing each page. Your responses for that page will not be saved until you click "Next". If you exit the survey before completing all pages, you can return at any time to complete it by returning to the URL. You will be directed to the last unanswered page. You can navigate between questions until completion. If you change you response on a previous page, those changes will not be saved unless you click "Next". Click "Next" to answer the pre-survey questions.

Pre-Survey Questions:
1. Name of your organization:
2. Length of time with your organization (in years):
3. Please select your organization type. (Federal, State, Tribe, University, Extension, Private, NGO, Local/County/Regional)
4. Please select your position type. (Manager, Educator, Researcher, Staff, Director)
5. In what area is your academic training? (Check all that apply. Biological Science, Environmental Science, Physical Science, Engineering, Social Science, Planning, Business, Economics, Public Health, Other)
6. List your area(s) of expertise. (e.g. agriculture, utilities, vegetation, water fisheries, etc.) Please list all.
7. List the geographic area(s) within your expertise (e.g. entire NW, Idaho, Oregon coast, Puget Sound, etc.)
8. What is your personal level of concern about climate change? (Very concerned, Moderately concerned, Not very concerned, Not concerned at all, Don't Know)
9. How well informed do you fell about climate change? (Well informed, Moderately informed, Not well informed, Not at all informed, NA)

Risk Questions:
1. Within your area(s) of expertise, please list populations/areas vulnerable to this climate risk.
2. Within your area(s) of expertise, please list important consequences were this risk to occur.
3. To the best of your knowledge, how would you rate the consequences of this risk in the following criteria: Human Health & Welfare: Human Population Affected (Few, About Half, Most, Don't Know, NA; Please provide rationale (optional))
4. Human Health & Welfare: Human Mortality (Decrease in mortality, Sign Uncertain, Increase in Mortality, Don't Know, NA; Please provide rationale (optional))

5. Human Health & Welfare: Health Quality (Improved general health quality, sign uncertain, reduced general health quality, extreme reduction in general health quality, Don't Know, NA; Please provide rationale (optional))
6. To the best of your knowledge, how would you rate the consequences of this risk in the following criteria: Natural Environment: Biodiversity (Favorable for increased biodiversity, Little or no change, Unfavorable for biodiversity, Severe reduction in biodiversity, Don't Know, NA; Please provide rational (optional))
7. Natural Environment: Geographic area affected (Few isolated areas, Moderate portion of the region, Most of the region, Don't Know, NA; Please provide rational (optional))
8. To the best of your knowledge, how would you rate the consequences of this risk in the following criteria: Built Environment: Public Services & Infrastructure (Little Disruption, Moderate Disruption, Major Disruption, Don't Know, NA; Please provide rationale (optional))
9. To the best of your knowledge, how would you rate the consequences of this risk in the following criteria: Economy: Cost (e.g. revenue loss, repair costs) (Possible net economic gain, No change/Sign uncertain, Low net economic loss – up to several million, Medium net economic loss – up to ~$50 Million, High net economic loss – more than ~$50 Million, Don't Know, NA; Please provide rationale (optional))
10. Is there any aspect of the climate risk and likelihood summary that the presenter has just provided that you would modify or add? If so, please briefly comment.

For a complete report of the survey responses and results, please contact Meghan Dalton, mdalton@coas.oregonstate.edu.

Workshop Sponsored by:

Climate Impacts Research Consortium (CIRC)
The Pacific Northwest NOAA-funded RISA
http://pnwclimate.org

Northwest Climate Science Center (NW CSC)
funded by US Department of the Interior

USGS Grant

Information about the National Climate Assessment
and the US Global Change Research Program can be found here:

http://www.globalchange.gov/what-we-do/assessment

Oregon Climate Change Research Institute
Oregon State University
(541) 737-5705
http://occri.net

Northwest National Climate Assessment Risk Framing Workshop

Friday, December 2, 2011
8:00 a.m. to 5:00 p.m.

DoubleTree by Hilton
1000 NE Multnomah Street
Portland, OR 97232
(503) 281-6111

8:00-8:30 am	Check-In & Continental Breakfast – Oregon Room
8:30-8:50 am	Welcome & Overview of the Day Philip Mote, Director, Oregon Climate Change Research Institute
8:50-9:30 am	Incorporating Risk Management into the Northwest NCA Meg Jones, Office of the WA State Insurance Commissioner and Gary Yohe, Vice-Chair, NCA Development Advisory Committee (remote)
9:30-10:20 am (5 minute presentation followed by 5 minute survey)	Climate Risk Panel Presentation & Audience Survey 1. Extreme heat events (Phil Mote) *Increases in average annual air temperature and likelihood of extreme heat events.* 2. Hydrology & water supply (Dennis Lettenmaier) *Changes in hydrology and water supply; reduced snowpack and water availability in some basins; changes in water quality and timing of water availability; including increased incidence of drought.* 3. Wildfires (Jeremy Littell) *Increases in wildfire frequency, intensity, and severity.* 4. Ocean temperature & chemistry changes (Kathie Dello) *Increases in ocean temperatures, with potential for changes in ocean chemistry and increased ocean acidification.*
10:20-10:40am	Morning Break (Coffee & Tea provided)
10:40-11:30 am	Climate Risk Panel Presentation & Audience Survey (cont'd) 5. Sea level rise & coastal hazards (Peter Ruggiero-remote) *Increased coastal erosion and risk of inundation from increasing sea levels and increasing wave heights and storm surges.* 6. Shifts in distributions of plants & animals (Sarah Shafer) *Changes in abundance and geographical distributions of plant species and habitats for aquatic and terrestrial wildlife.* 7. Diseases and pest outbreaks (Jeff Hicke –remote) *Increases in diseases and insect, animal and plant pests.* 8. Extreme precipitation & flooding (Phil Mote) *Increased frequency of extreme precipitation events and incidence and magnitude of damaging floods.*

11:30-11:40 am	Overview of afternoon breakout sessions
11:40-12:40 pm	Lunch (provided) – Idaho/Alaska Room

12:40-1:50 pm	Breakout Session I – Oregon Room
	Identify & rank consequences
	    a. Extreme Heat Events
	    b. Changes in hydrology & water supply
	    c. Wildfires
	    d. Ocean temperature & chemistry changes

1:50-2:20 pm	Report to Plenary I
2:20-2:40 pm	Afternoon break (Coffee & Tea provided)

2:40-3:50 pm	Breakout Session II – Oregon Room
	Identify & rank consequences
	    a. Sea level rise & coastal hazards
	    b. Shifts in geographic distribution of plants & animals
	    c. Spread of diseases, invasive species, and pest outbreaks
	    d. Extreme precipitation & flooding

3:50-4:20 pm	Report to Plenary II
4:20-4:50 pm	Plenary Discussion Discuss, defend, refine aggregated risk ranking
4:50-5:00 pm	Concluding Remarks
5:00 pm	Adjourn

APPENDIX D: EXAMPLE SURVEY RESULTS REPORT

Risk 1: Extreme Heat Events

Within your area(s) of expertise, please list populations/areas vulnerable to this climate risk.

1 children and adult with chronic disease , poor in vulnerable areas critical and already at risks species

2 Human populations in urbanized areas; vulnerable human populations in rural areas that do not have effective means to provide shelter. Ares of the state that depend in part on mountain snowpack for water supply.

3 1) fisheries - particularly those that are already endangered or threatened and those that have no where to move 2) agriculture/farming – increased temperatures would likely create need for increased irrigation, but with climate change, natural storage rights (in-stream use) may become less available. Stored water may also become less available in the PNW because their is likely going to be a greater draw on stored water due to decreased in-stream flow during the warmer season.

4 disadvantaged urban human populations rural populations near forest boundaries tree & agricultural species foraging fish, shallow water species alpine & sub-alpine plant & animal species

5 fish (esp salmonids), wetlalnds, streams and rivers, forests

6 Agriculture (particularly crops that need chill hours, etc.), forests (reduced likelihood of pest die-off), broad ecosystems - potential for advancement of invasive species that otherwise would be restricted by temp; cold-water fish (including higher water temps, altered streamflows if temperature changes alter snowpack and runoff patterns).

7 cold water fish populations; forests; fruit trees and vines; human populations in the inland PNW (E. WA, E. OR, S. Idaho); energy systems that deliver power to the western US grid in summer - particularly a big issue for heat waves in California where human population and cooling demand is greatest; risk for vulnerable human populations in inland NW to heat-related stress (elderly, poor, people in poor health, without access to air conditioning)

8 Trees species in the Pacific Northwest

9 Geographically isolated species and populations; cool and cold water fishes; species with obligate-level relations with geophysical AND climatic niche-spaces

10 Potential impacts to snow pack and hydrology, therefore affecting Columbia basin include the entire Northwest. Populations would include human, particular fish species, etc.

11 high altitude salmon habitat in the Cascades, snow-cover dependent species, upland species, tribal elders, aquatic species

12 salmon during summer low flow periods, esp in interior Columbia basin; typically vulnerable urban populations (poor, elderly); farm workers;

13 Older people, children, those with cardiovascular disease/compromise, some env justice communities, urban/suburban communities,

14 plant and wildlife populations are vulnerable

15 southern Idaho agricultural irrigation districts, mountain communities in forested areas, communities in range lands, urban areas around and including Boise

16 10 Million

17 Tribes, plant and animal species of cultural importance to tribes

18 Salmonids (particularly in streams and oligohaline portion of estuaries) Possibly commercial aquaculture (i.e., oysters) due to increased disease linked to warmer temperatures; Increase in harmful algal blooms (Stephanie Moore's research from Puget Sound).

19 Water treatment facilities.

20 Salmonids and other aquatic organisms in streams without cold water (i.e., groundwater) sources; urban areas (Portland, Seattle, but probably most likely Boise)

21 Higher likelihood of forest insect outbreaks with decrease in cold spells throughout PNW/Alaska. Increase in taiga fire frequency and loss of permafrost already observed. Increased fire frequency in

shrubsteppe vegetation.

22 Cold water fish and other aquatic species; species that are drought prone; increased competition from species, pathogens, pests that have historically been restricted in range due to winter lows

23 People, Salmonids, some crops

24 areas vulnerable: eastern OR, WA, S ID populations: forest species (e.g. lodgepole pine)

25 entire region and NW states, urban areas in particular

26 Shellfish aquaculture

27 Urban concentrations of people from rural to larger areas, water temperatures for trout and salmon elderly people and poor housing with no air conditioning relief

28 coastal, water sheds

29 Areas less adapted to hot weather, I.e the coast may experience more heat illness morbidity. These areas are less likely to have air conditioned homes. Elderly, the ill, and pregnant women, those who work outside, and lower income may be more vulnerable to extreme heat events.

30 Crops are vulnerable to heat stress, particularly during flowering, with potential for significant yield reduction

31 urban poor and elderly, heat-sensitive species

32 Outdoor workers Elderly Chronically ill Disabled

33 agricultural producers, coastal residents

34 Urban / populated areas Urban wildland interface (increased fire ignitions, higher fire spread potential given extreme fire weather)

35 ag ecosystems marine coastal areas

36 elderly, urban areas

37 all

Within your area(s) of expertise, please list important consequences were this risk to occur.

1 infrastructure damage, livestock, species dependent on cool water risk respiratory problem for human irrigated agriculture -- increase in water needs water loss crop yield reduction for dryland increase ocean acidification due to accelerated decomposition of organic matter

2 Drought; heat-related deaths; increased chance of wildfire.

3 1) Fisheries issue - extinction of fisheries is the highest consequence, which would have a domino affect on the ecosystem (any species relying on those that become extinct may be in trouble, etc); 2) Agriculture issue - this could cause food shortages causing prices to skyrocket (e.g., wheat/grain a few years ago).

4 increases in acute and long-term human health impacts; decreases in local food supplies; air quality impacts from forest fire; reductions in amount and quality of summer and fall water supplies;

5 decrease snowpack, changes stream hydrology, areas refugia, soil moisture changes, decrease wetlands

6 Consequences: loss in agricultural revenues; increased potential for forest disturbances; declines if cold water fish populations

7 large financial losses for fruit and possibly wine growers; forests at risk to extreme drought stress during prolonged heat waves; summertime migrating salmon rearing or resident cold water fish at risk of major fish kills during hot spells that warm water temps above ecologically significant thresholds; potential for energy grid failures at periods of exceptionally high demand could lead to rolling blackouts

8 Local extirpation or extinction of tree species could result.This could change the structure and function of ecosystems in the Pacific Northwest. Economic, social, and ecological instability could result. Forests and tree species contribute substantially to the northwest economy through timber and other products; they also offer water quality control, flood protection, fisheries protection, carbon storage for climate change mitigation, and many other products and services.

9 Disruption of population structure, popuulation and community fragmentation; population and species loss

10 Changes in snowpack, runoff timing, availability of water for multiple uses including for hydropower

generation, flood control, fish and wildlife needs, irrigation, etc

11 lethal temperatures for some aquatic species, less likely to be lethal to elders in the Puget Sound, loss of infiltration capacity of soils with prolonged drought/heat waves

12 increased human morbidity and mortality - positive interactions w/ respiratory ailments; barriers to salmon migration, altered timing of salmon migrations, increased salmon mortality

13 Increased mortality and morbidity, socio-economic costs of expanded aqm programs,

14 changes could occur to the competitive abilities of plant species thus changes in plant species composition in areas most affected would also occur. Changes in plant species will, in turn, affect wildlife and invertebrate species dependent on the plant species that have experienced stress or loss. Economic expectations from forest management could also be affected because primary production would be altered among tree species.

15 increased demand on water supplies, greater likelihood of fire

16 Impacts to culturally important species from increased warming; shift in growing seasons, access to species for subsistence and economic needs.

17 Decline in salmonids, possible decline in water quality in streams & oligohaline portions of estuaries (e.g., decline in oxygen levels due to increased respiration & reduced saturation)

18 Issues with water availability and water quality.

19 Shift towards warmer-water species in streams, likely increase in stream-and lake borne pathogens (i.e., toxic algae blooms); altered life history timing (i.e., migration, spawning)

20 Tree mortality and conversion of ecosysten dominant tree species, hence, biotas. Conversion of shrubsteppe to exotic-annual grassland. Increased loss of carbon to atmosphere through fire and loss of organic peats in taiga. Change in dominant taiga forest species from coniferous to deciduous.

21 species range shifts; increased stressors on species and habitats; ecosystems with new assemblages of species-- unknown results of interactions, etc.; increased risk of fire, increased fire frequency; increased risk of pathogens, disease, and competition or toher consequences from increased native (pest) and non-native species

22 People: some additional heat related deaths; salmon-decline and or extinction based on poor rearing conditions, cool-temperature crops - poor yields of unable to grow; some crops prmaturely exposed to fost damage or failure to harden over winter; Highe/peaky demand for electricity for A/C

23 forest mortality (already occurring)

24 Increased peaking needs of energy load to meet additional cooling needs will lead to more need for additional energy generation resources, and likely increased greenhouse gas emissions ... Potential impacts on wind power generation although precise impacts are difficult to ascertain. Potential impacts on materials from extreme heat, concrete especially. Perhaps shorten lifetime of dams and such.

25 Increased cases of pathogenic Vibrio and potentially toxic red tide events -- relationships with air temperature are strong, but note that to my knowledge impacts of extreme heat events haven't been evaulated (I imagine for Vibrio the risk is high)

26 Fish risk related to regulatory requirements Need to meet peak day water supply needs in summer Flooding risks for infrastructure facilities

27 changes in water flow to estuary changes in sea surface temperature changes in water stratification and delivery of nutrients to surface waters

28 Increased morbidity and mortality related to heat illness. Extreme heat can also affect air quality- so could increase bad air days and affect respiratory health and cardiovascular events.

29 More important than number or duration of events, it is the absolute Tmax that will take place. For example, wheat and maize (and other cereals in the region) will experience yield reductions with exposure to temperatures above 30 C.

30 heat stress, heat-related morbidity, species range loss for heat-intolerant species, expansion of cold-intolerant species esp pests

31 Heat-related illness (heat rash, heat syncope, heat exhaustion, heat stroke)

32 changes in income, risk of damage from sea level rise

33 Increased rate of wildfire ignitions and extreme fire weather translate to larger fires, more fires, potentially more extreme fires that are harder to suppress, and when they occur near populated areas, the impacts on infrastructure, forests, and people can be large through fire and air pollution (smoke, PM2.5

particulate)
34 reduce yields changes in land use patterns
35 human morbidity and mortality, increased vectorborne disease
36 changes in snowmelt runoff (associated with warmer temps) and annual flows -- obviously this is an indirect, rather than direct effect. there may be some indirect effect also associated with increased evaporative demand.

Human Health & Welfare:

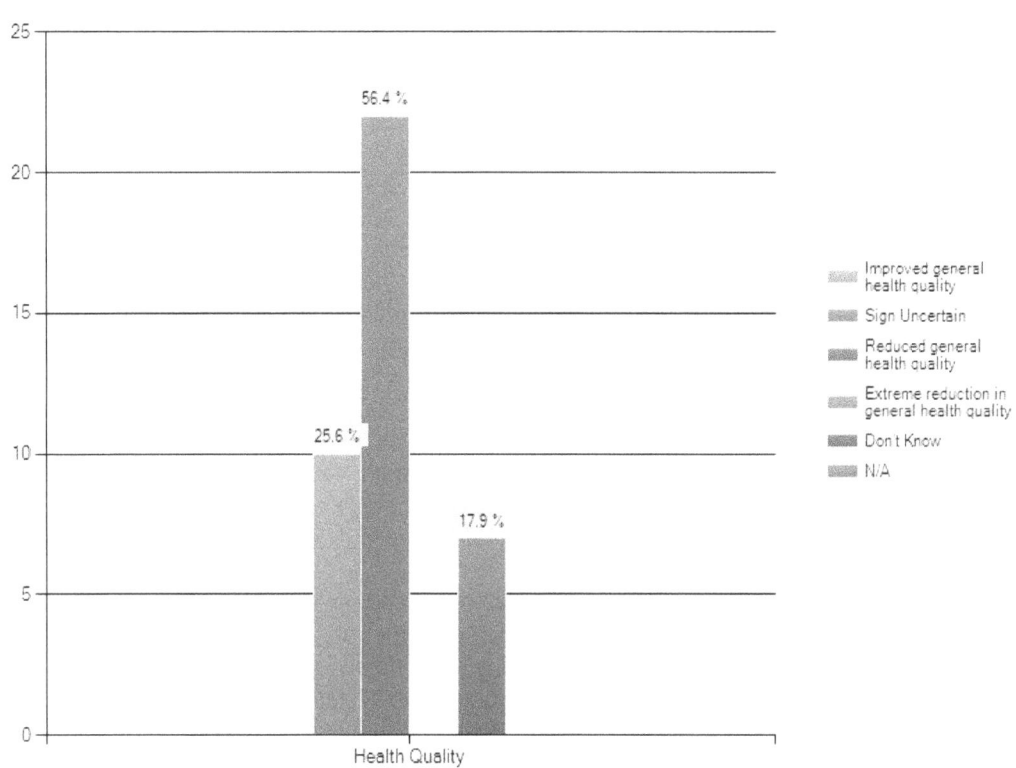

Human Health & Welfare:

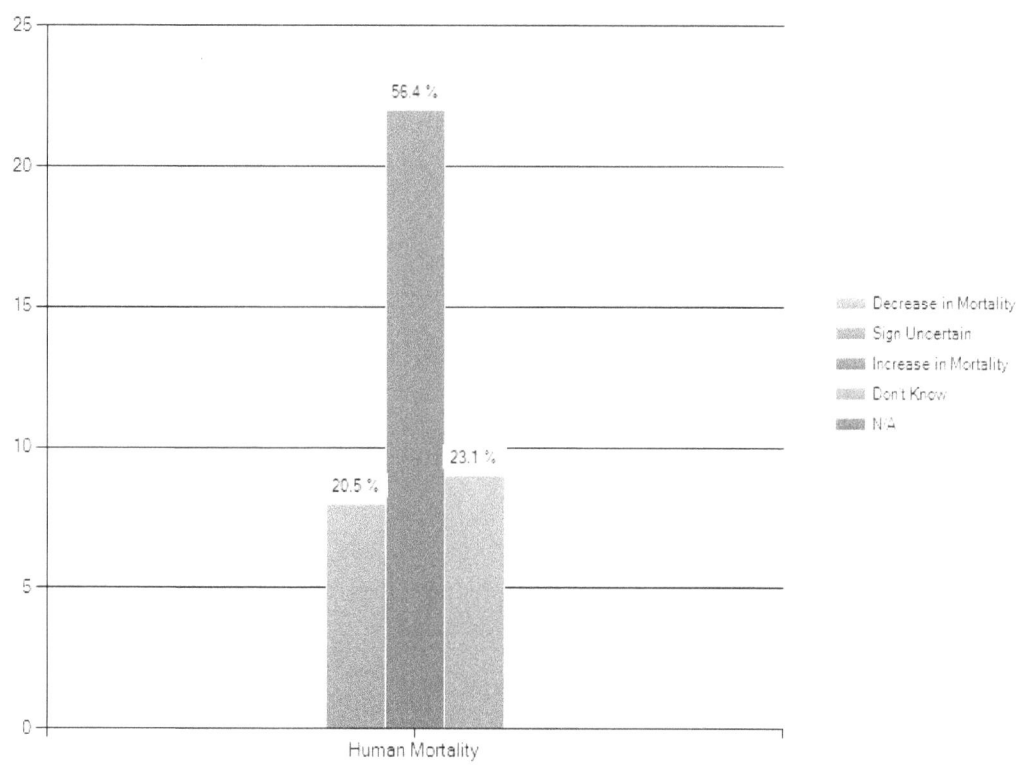

To the best of your knowledge, how would you rate the consequences of this risk in the following criteria: Natural Environment:

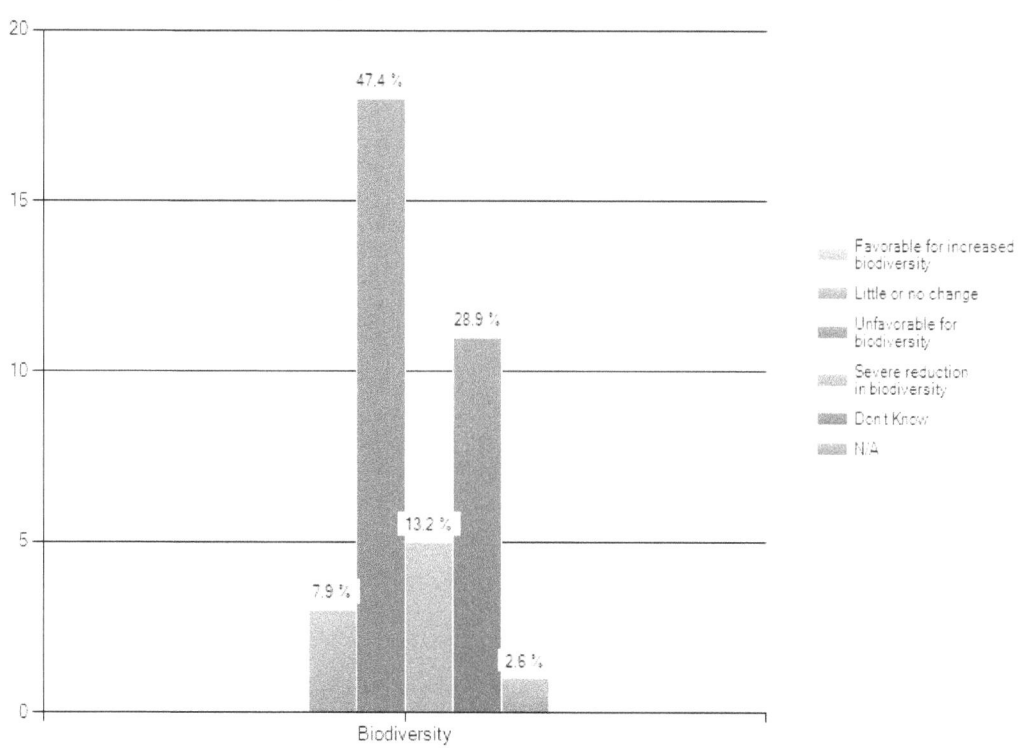

Natural Environment:

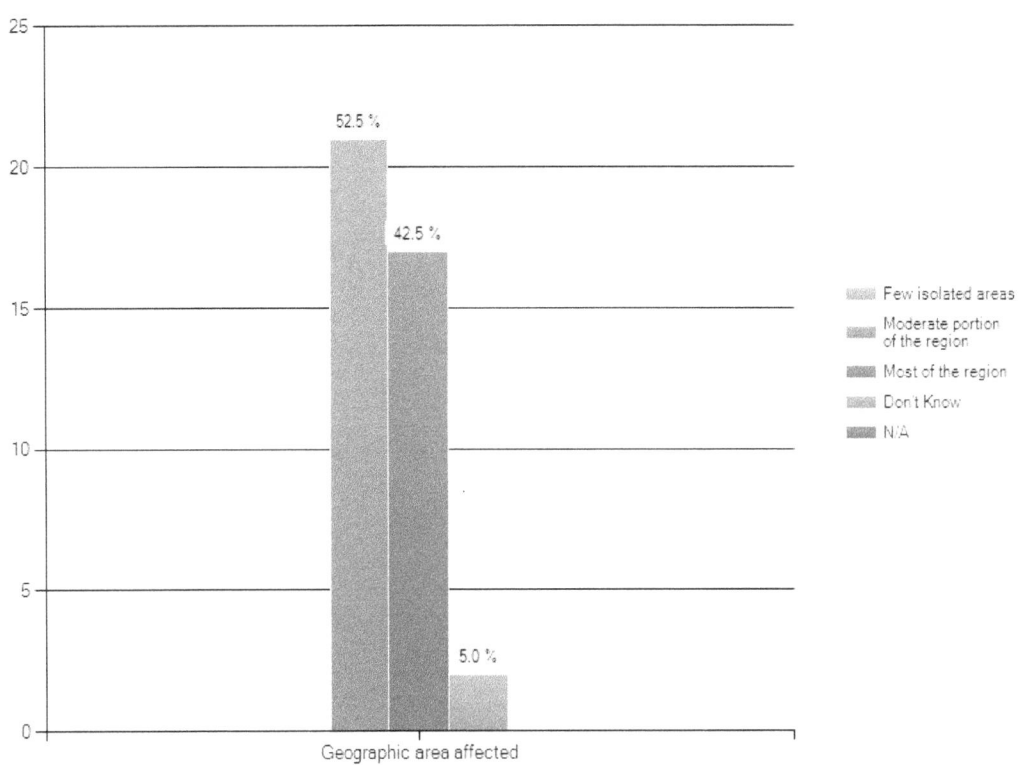

To the best of your knowledge, how would you rate the consequences of
this risk in the following criteria: Built Environment:

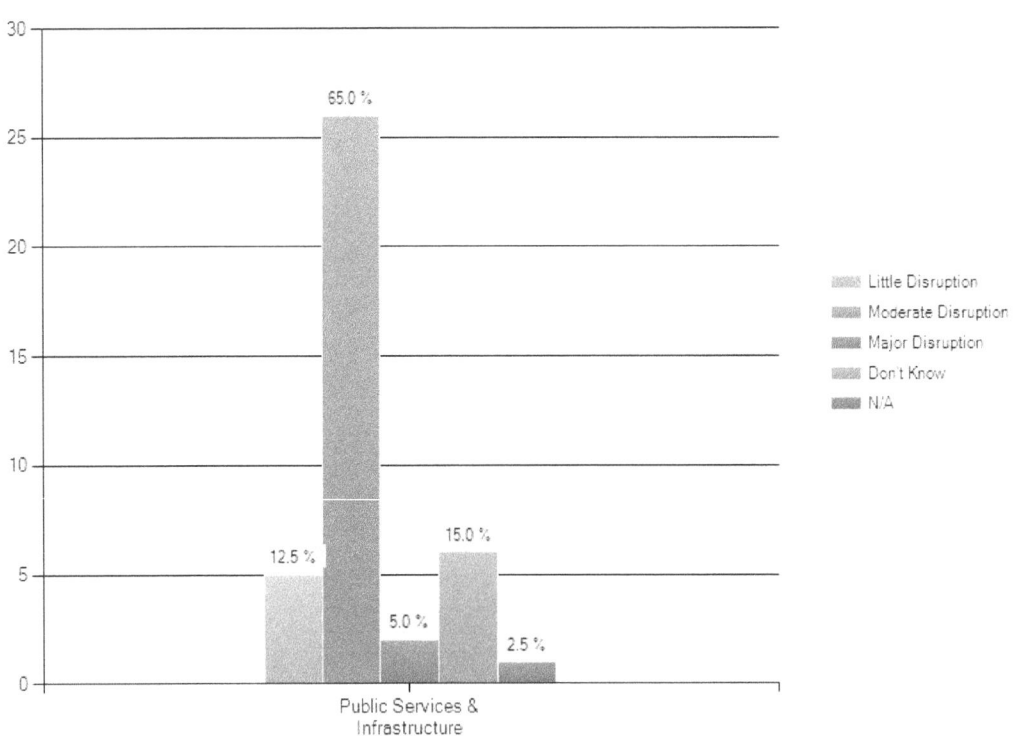

To the best of your knowledge, how would you rate the consequences of this risk in the following criteria: Economy:

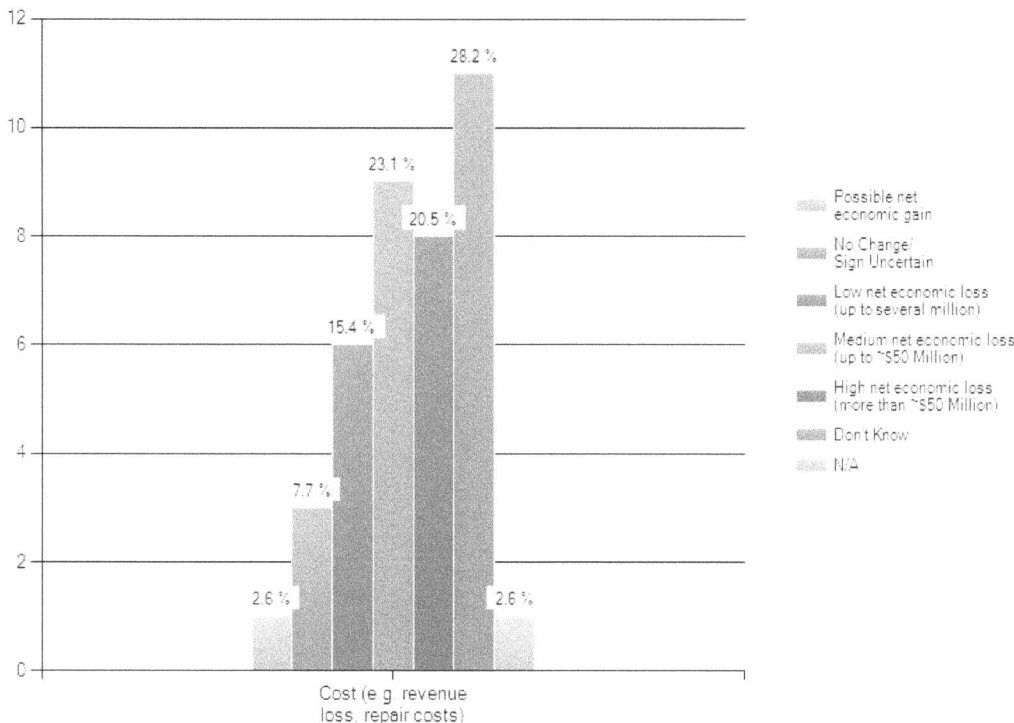

Cost (e.g. revenue loss, repair costs)

Legend:
- Possible net economic gain
- No Change/ Sign Uncertain
- Low net economic loss (up to several million)
- Medium net economic loss (up to ~$50 Million)
- High net economic loss (more than ~$50 Million)
- Don't Know
- N/A

To the best of your knowledge, how would you rate the consequences of this risk in the following criteria: Human Health & Welfare:

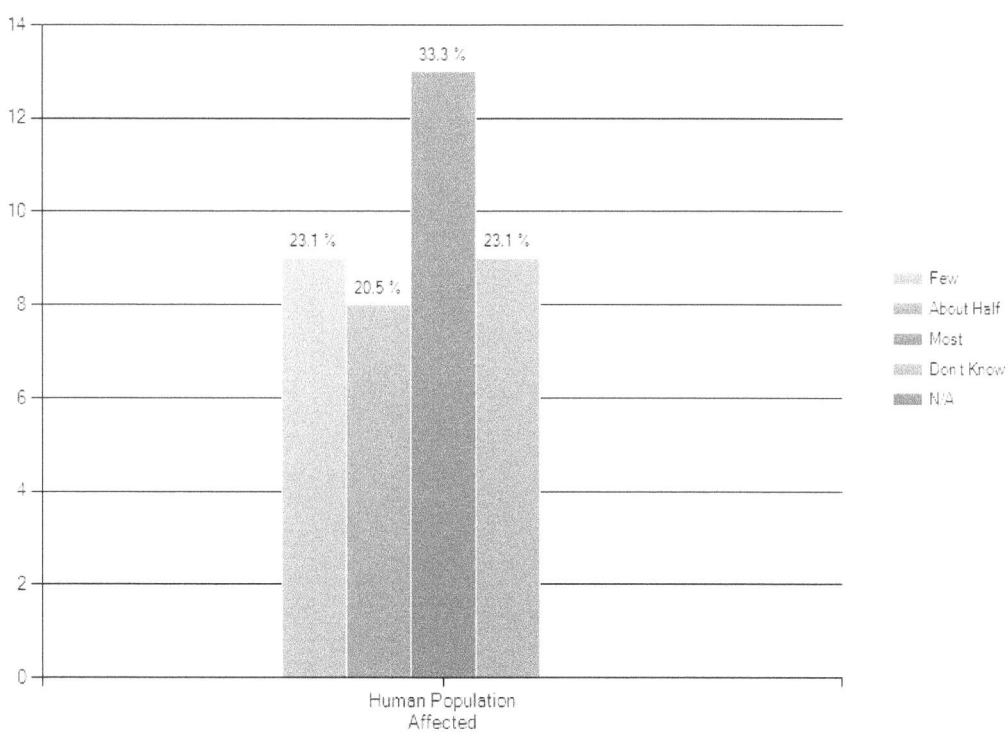

Human Population Affected

Legend:
- Few
- About Half
- Most
- Don't Know
- N/A